P9-DTZ-991

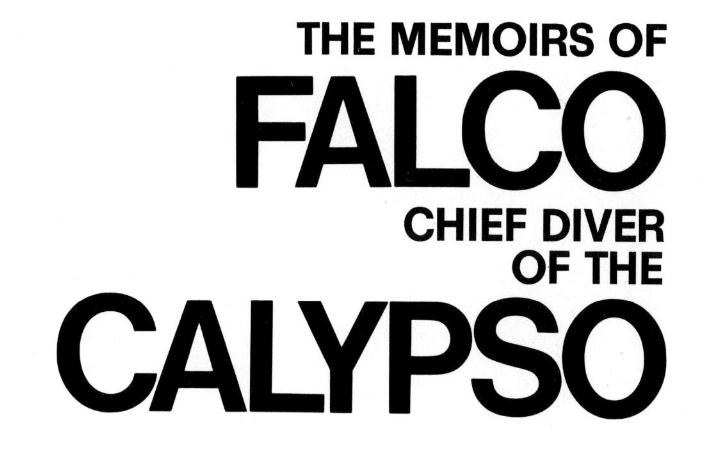
THE MEMOIRS OF
FALCO
CHIEF DIVER
OF THE
CALYPSO

THE MEMOIRS OF FALCO CHIEF DIVER OF THE CALYPSO

by

PHILIPPE DIOLE
and ALBERT FALCO

Translated from the French by
JOSEPH HARRISS

BARRON'S
WOODBURY / NEW YORK

© Copyright 1977 by Barron's Educational Series, Inc.

© Flammarion, 1976

All rights reserved.
No part of this book may be reproduced
in any form, by photostat, microfilm, xerography,
or any other means, or incorporated into any
information retrieval system, electronic or
mechanical, without the written permission
of the copyright owner.

All inquiries should be addressed to:
Barron's Educational Series, Inc.
113 Crossways Park Drive
Woodbury, New York 11797

Library of Congress Catalog Card No. 77-376

International Standard Book No. 0-8120-5130-0

PRINTED IN FRANCE

CONTENTS

422

preface

A thousand years from now, if the human race survives all sorts of dangers that threaten it, Albert Falco—Bébert—will be mentioned in books as one of the first two "oceanauts"; that is, one of the pioneers who lived for the first time in an "underwater house" (Conshelf I, September 1962). Falco the oceanaut will have the same status as Gagarin the cosmonaut, for he will have blazed the path to modern industrial saturation diving.

What historians will not be able to evaluate, however, are the influence Bébert has had on those who know him, the affectionate esteem of his fellow workers, and the mass of extraordinary achievements he has accomplished with disarming modesty.

Although he is 20 years younger than I, Albert Falco has assumed as much importance in my life as the two other fine men with whom I began my exploration of the seas—Philippe Tailliez and Frédéric Dumas. Philippe taught me respect

Commander Cousteau helps Albert Falco into his diving suit.

for life, the unity of all things, the importance of following one's dream. Didi taught me to observe, to economize my strength, and to move with ease in three dimensions. As for Bébert, he literally belongs to the sea. Aboard the Calypso, *he has spent 22 years helping to organize and run our expeditions as director of diving operations; even during his vacations his favorite form of relaxation is diving! Endowed with a sharp intuition and a love for all living things, Bébert has acquired a reputation in the scientific world as a remarkable observer of the behavior of marine animals. When it is impossible for me to dive in an area I would like to explore, I depend on Falco to be my eyes.*

One summer day in 1952, the Calypso *was moored a few yards off a small island known as the "Grand Congloué," ten miles from Marseilles. For several months we had been working an underwater archeological site 135 feet down the base of the island, a site that was to enrich the Marseilles Museum with thousands of amphorae or other forms of pottery and numerous fragments of a ship dating from the third century* B.C. *Directed by Professor Fernand Benoit, the work was enormous and lasted several years. That day, J. Borély, founder and president of the first scuba diving club in Marseilles and a pioneer in underwater photography, came to see me on the* Calypso. *Unfortunately, the wind became so strong toward the end of morning that our mooring could no longer hold. The* Calypso *had to make for Marseilles, and Borély invited me to go with him by car.*

Along the way he stopped atop the desolate and magnificent cliffs of Marseilles-Veyre. The wind, known in Provence as the mistral, *was raging. Far below, the sea was churned to white and a tiny* Calypso *was struggling along, disappearing momentarily beneath clouds of foam thrown up by her bow. Moved by the power of the spectacle before us, Borély took me by the hand and shouted through the storm, "What a sight! And what a future we have before us! I'm going to make you the finest gift I can think of. I'm going to send you a certain young man I know."*

The next day, Albert Falco came aboard the Calypso. *He soon became chief diver, pilot of the "diving saucer," oceanaut, and my friend.*

Jacques-Yves Cousteau

1

Sormiou Cove

BENEATH THE SEA—MEETING TARZAN—A DANGEROUS OUTFIT—
THE WAR—NEGOTIATING THE RAIN AND WIND—MINE
CLEARANCE—TRIPS—THE FIRST GOOD SCUBA

Marseilles is surrounded by high white and gray cliffs in which the sea has hollowed out bays and grottos and small coves known locally as *calanques*. They make a wild, majestic landscape, with dizzying peaks towering hundreds of feet above the sea. Some of these inlets are accessible only by boat. They have long been miniature underwater paradises rich in fish and all sorts of other marine life. It was in one such cove that Falco spent his childhood, and that first contact with the sea has probably left its mark on his entire life.

He was born in Marseilles on October 17, 1927, but it was in Sormiou Cove that he took his first steps in life.

Sormiou means "good spring" in the Provencal language of the south of France. There is, in fact, a spring there—probably an ancient watering place for Greek and Phoenician ships.

Albert's parents owned a small hut at the inlet. You have to be from the south of France to understand what a hut like that means to the people of Prov-

Albert Falco and Commander J.Y. Cousteau aboard the *Calypso*.

Right, above: Albert Falco diving among gorgonia.

Right, below: Among the reefs of the Bahamas.

ence. Often it is no more than a wooden shack, but it is nearly always at water's edge, surrounded by boulders and pines. The people of Marseilles and Toulon "head for the hut" on holidays to drink anis-flavored *pastis* and eat fish soup and *bouillabaisse.*

Sormiou is perhaps the most beautiful and certainly the most pleasant of the five or six coves neighboring Marseilles. Pine trees running down to the water make shade in summer and cut off the wind in winter. But in those days there was no road to the sea. All food and other things necessary for a weekend in the hut had to be carried or strapped on a donkey, and the trail down to the cove was called "the donkey track."

"I wasn't quite a year old when my father took me to Sormiou," Falco recalls. "From Marseilles we had to take a trolley that went to the village of Mazargues. Then we took a little bus that left us at the foot of the hill. My father carried me on his shoulders. The climb up the path lasted 20 minutes. We had to pick our way over heaps of stones and scale large boulders, with me balanced precariously on my father's shoulders. The path twisted among the boulders. After a quarter of an hour we came to a place called '13 corners.' From there you could see the sea for the first time way down below, set like a blue stone among the cliffs. It's an image that has been very important for me. All during my childhood I experienced the vision of that intense blue suddenly appearing to me from the top of the hill like a sort of revelation or promise.

"From there, we had only one thought: get down to the sea, dip our feet into it, touch it. The bay was no longer a blue stone but living water, rich in innumerable lives, minuscule lives that a child could comprehend. It was water that invited me to enter.

"The hut represented not only vacations beside the sea, but also a certain atmosphere of family affection. Other images come to mind: the gleam of the oil lamp in the evening, the wood fire where fish were being grilled, the table where we took our meals, and, almost within reach, the water, smooth as a mirror or shivering with a slight lap.

"My most vivid childhood memories are of fishing trips we took out of the bay. My father had been in the navy during the First World War, serving aboard a mine sweeper. It was a strange coincidence that I was to spend most of my life aboard another mine sweeper, the *Calypso.* My father owned a little fishing boat with an outboard motor, and he began taking me with him before I started to talk.

"At Sormiou the sea fascinated me with its inviting and mysterious limpidity. I was attracted both by what I could and could not see beneath its surface. From the age of three I began to scoop at the water, making it spurt and sparkle in showers of light beneath the sun. I didn't know how to swim, or even to float, but I put my head under the water and even moved submerged for short distances. My mother was scared to death as she watched me.

"Soon I began to catch little fish along the shoreline, white and gray gobies that I forced up onto the beach with my hands. I put them in a little aquarium that my father built.

"I caught shrimp, too—there were lots of them in those days—in holes in the rocks. I discovered that white attracted them. Sometimes they swam into my hand and tickled it, which I thought was great fun."

Beneath the sea

If Falco is today one of the men who knows best the sea and the animals that live in it, if he has an intuitive feel and an almost physical understanding of it, it is because as a child he became aware that the sea was not, as those who live only on land believe, a closed, impenetrable surface, but an accessible substance where there were living beings. Just consider the importance of such a revelation for a child three or four years old.

"I spent my time in the water or on a little pier," he says. "I explored bit by bit the rocks bordering the bay.Most of the time I threw back the fish I caught, but I had satisfied my desire to see them, to approach them, to touch them, whereas in the sea they always disappeared too fast. I think that it is thanks to these childish activities that I learned little by little to guess the reactions of fish and to know how they would act—a way of dominating them. I discovered that they were neither inaccessible nor stupid. For years I watched them live. I came close to fish that were bigger and bigger and also faster in the water than me.

"One day when I was about five, I saw an enormous conger eel in a hole on the bottom. I ran and asked my father for a hook. He gave me a big one that he used to fish for tuna. I baited it with sea snails and dangled it in front of the eel, which gulped it down. It was a really big one, bigger than me. When the eel pulled on the line I fell in and got tangled up with it. We thrashed around in the water while several persons who had seen me fall in called my father. He came running and pulled us both out. Luckily I hadn't been bitten."

Sormiou Cove is bounded on both sides by sheer cliffs riddled with holes and grottos. On the side facing the open sea there are large underwater vaults and corridors. As a youth Albert Falco explored this miniature world year by year. He

Albert Falco in the Capellans Grotto, Sormiou Cove.

At right: Near the same grotto, 90 feet down, the marvelous beauty of yellow gorgonia, madrepores, and ascidium.

filled his lungs with air, dived, and made many observations: in January and February, for example, female sea perch came to lay their eggs in one of the grottos. In June, spider crabs attached themselves to the cliff face. In May, octopi came out of their holes. Falco thus learned the seasons of the sea.

As soon as he learned to swim, at about four or five, he began to visit the other coves around Sormiou, like Petit Soldat and Santiago. Some of them were accessible only during the summer because in winter the sea was too rough. But Falco learned not to fear waves. As he would be caught by a foaming giant that was carrying him toward a sharp rock, he would become adept at riding the wave so it would carry him over the rock. He played with the sea.

The shoreline around Marseilles is composed of these high chalk cliffs towering above the water, giving the area a magnificent if fearful landscape. The entire coastline is made of stone, sometimes smooth and sometimes smashed to bits as if by a cataclysm, and often vertiginously high. This somewhat chaotic site was also the domain of Falco and those of his friends bold enough to follow him. When he was not in the water he went in for mountain climbing.

Constantly pounded by the sea, those gray and white rocks seem to be devoid of any plant life. But Falco learned otherwise.

"Children like us," he says, "used to climbing all over those cliffs knew that they held many good things that varied with the seasons. First of all we found herbs like thyme and rosemary. In January and February we gathered wild asparagus and lettuce for salads. In the fall there were briar and mushrooms. We discovered that life can exist just about everywhere. That's one lesson that I remembered later, when research cruises took me to the Arctic, the Antarctic, or Tierra del Fuego—regions that most people believe to be completely barren. But life can sprout everywhere, and flowers bloom even in a small hole in solid rock. No matter how wretched a place may be, life will find a foothold."

Meeting Tarzan

When he was 12, Albert Falco lost his father. His life was seriously upset by his no longer being able to go fishing or spend his days at sea. He was alone, and his mother forbade him to go to sea by himself. He was utterly at loose ends.

One day as he was sitting sadly on the beach at Sormiou, he saw a man loading a strange gadget into a canoe. It was a sort of lance about ten feet long, with a blade at the end. For several days Falco watched with rising curiosity as the man returned to shore with 10 or 15 fish. Then one day the man asked, ''Would you like to come along?''

His name was Georges Beuchat, and he was 15 or 16 years older than Albert. In love with the sea, he learned from a Tahitian named Canaldo how to spear fish. Falco was fascinated by Beuchat, but he was still only a child. He had neither equipment nor boat nor money. He especially would have liked to be able to look underwater like Beuchat, who used Fernez industrial goggles. Spearfishing was just beginning. It was to be the origin of the first underwater exploration.

When he began going to sea with Beuchat, Falco finally was able to leave the cove and head for the deep water beyond the towering cliffs. The shallow waters of his childhood were behind him.

Georges Beuchat lent him the goggles.

''I entered the water holding onto the canoe with one hand,'' he remembers, ''because I was afraid of those blue depths, of that void that gave me a feeling of vertigo. It was in Cancéou Cove. I watched Beuchat dive deeper ahead of me with his long harpoon. For me, Beuchat was Tarzan. A school of grouper was drifting in front of a hole in the rock. There were five or six, including two large ones, and he speared the biggest. I was divided between fear and curiosity. The fish was powerful and put up a good fight, which I watched despite the emotion that made me shiver in the water. At last Beuchat surfaced with his prize. The whole incident seemed extremely long to me, and I thought to myself that I would never be able to do that.

''From then on we were great friends. He took me out every day in his boat. I was a good rower, able to pull on the oars for hours. First he gave me a pair of goggles and then a snorkel that I used with a nose clip. Bit by bit I got to know the sea better by looking into it and learning to move in it.''

But we also have Beuchat's recollection of what Falco was like at that time:

''When I saw him dive from the boat at 13, he was already so well adapted to the sea that he hardly made a splash. And when he swam underwater he had an eagle eye.''

Using goggles, Falco was finally able to see in the ocean what he had previously only guessed at. It was like re-discovering Sormiou Cove and its environs. The sandy bottom sloped down gently at first, then came a bed of seaweed that covered the center of the cove to a depth of about 90 feet. Along the shoreline were masses of broken boulders that formed a wall that fell away sharply toward the mouth of the cove. There were innumerable fish, including mullet, sars, and especially grouper.

Falco was 14 when Beuchat finally gave him a complete underwater fishing

Albert Falco brings back an amphora fragment he has found on the bottom of Sormiou Cove.

At right: One of the red gorgonia that grow on the underwater cliffs of Cape Morgiou.

outfit with a long harpoon with detachable head. Beuchat himself made the harpoon out of three metal tubes totaling about nine feet long; they were shot with a thick rubber band.

On one occasion Falco was swimming with his harpoon when he came up against an enormous grouper. He shot. A few moments later he was left holding a twisted harpoon without a head. The grouper was gone.

Falco recalls that in those days there were at least 300 grouper in the cove. Every hole in the rock seemed to have its fish, and they were not afraid of men. Indeed, they were easy to see up close, and appeared to stand on their tails while waiting for a visitor to descend from the surface.

A dangerous outfit

Georges Beuchat, Albert Falco, and their friends loved spearfishing, but what they liked best was deep diving. Their dream was to be able to live underwater without suffocating by spending a couple of minutes with their respiration blocked. Whenever they heard of some amateurish invention that was supposed to help someone dive better, they were ready to try it, whatever the risk.

One day Beuchat came to Sormiou with an engineer from Marignane who had a scuba outfit he had invented. They tried it out in the middle of the cove. Beuchat, who tried it first, began to suffocate and nearly drowned.

"Go on," he said to Falco, "give it a try. Go down to 45 feet, near the bottom. You'll see, the air comes through better down there."

It was better, but it was far from perfect. Falco too almost stayed down for good.

There was another try near the island of Riou. This time Beuchat had brought along an English scuba outfit made by Siebe and Gorman with a copper helmet. But for such novices the gear was not easy to use. These experiments disappointed the group interested in scuba diving.

The war

Then came the war, with the occupation of the coast first by the Italians and then by the Germans. It became very difficult to go spearfishing because private boats were not allowed out of port.

But things worked out—they always seem to in the south of France, even under German occupation. Sormiou Cove was guarded by a German detachment. Falco discovered that its commander was an athlete who loved the sea. He showed him how he dived with his breath blocked and obtained permission to go out to sea.

But the Germans had taken the outboard motor off Falco's boat. He had to row for two or three hours to get to the island of Riou and the cliffs of Devinson, near Morgiou, where the chalky walls rise nearly 100 feet. They are said to be the highest in Europe. Here the water is crystal clear above a bottom of large boulders. It is a virtual paradise, a gigantic aquarium filled with fish.

One day Falco and Beuchat were rowing toward the open sea and were stopped in the middle of the cove by a burst of machine gun fire. There was a new commander of the German detachment, and he refused them permission to go out. So they found other ways to get to the sea. They walked along the cliffs as far as Cancéou or Morgiou and then descended ropes to reach the water.

It was also forbidden in Marseilles to go fishing, but it became increasingly difficult to obtain food, and fishing became a vital necessity. Falco and his friends looked all along the coast for outlets to the sea that were not guarded by sentinels.

"From Marseilles to La Ciotat," says Falco, "I know all the bottoms, all the cliffs, stone by stone and grotto by grotto."

At certain strategic points the Germans had put underwater nets and had posted sentinels. Falco managed nonetheless to get through these barriers by swimming under the nets when the guards were not looking. But occasionally the Germans did see him, shouted for him to stop, and fired rifle shots in his direction. Those were awkward moments.

Since it was impossible to row out of Sormiou, the group took the train or rode bicycles to Carry-le-Rouet or to La Redonne, where the Germans were less vigilant. It was at La Redonne that their adventure almost went bad. The sea was completely blocked off by German guards. Falco and his friends went through a railway tunnel toward Marseilles and came out on the shore opposite an island called Evelyne. There they entered the water armed with new device made by Beuchat, a wooden speargun.

When Falco was swimming back to the shore he saw a large grouper under a rock and shot a spear from the speargun. He hit it well, but it took three more spears to finish off the great fish. Falco was now 16, and it was his first big catch.

Double page following: Aerial view of Sormiou Cove. Photo Sciarli.

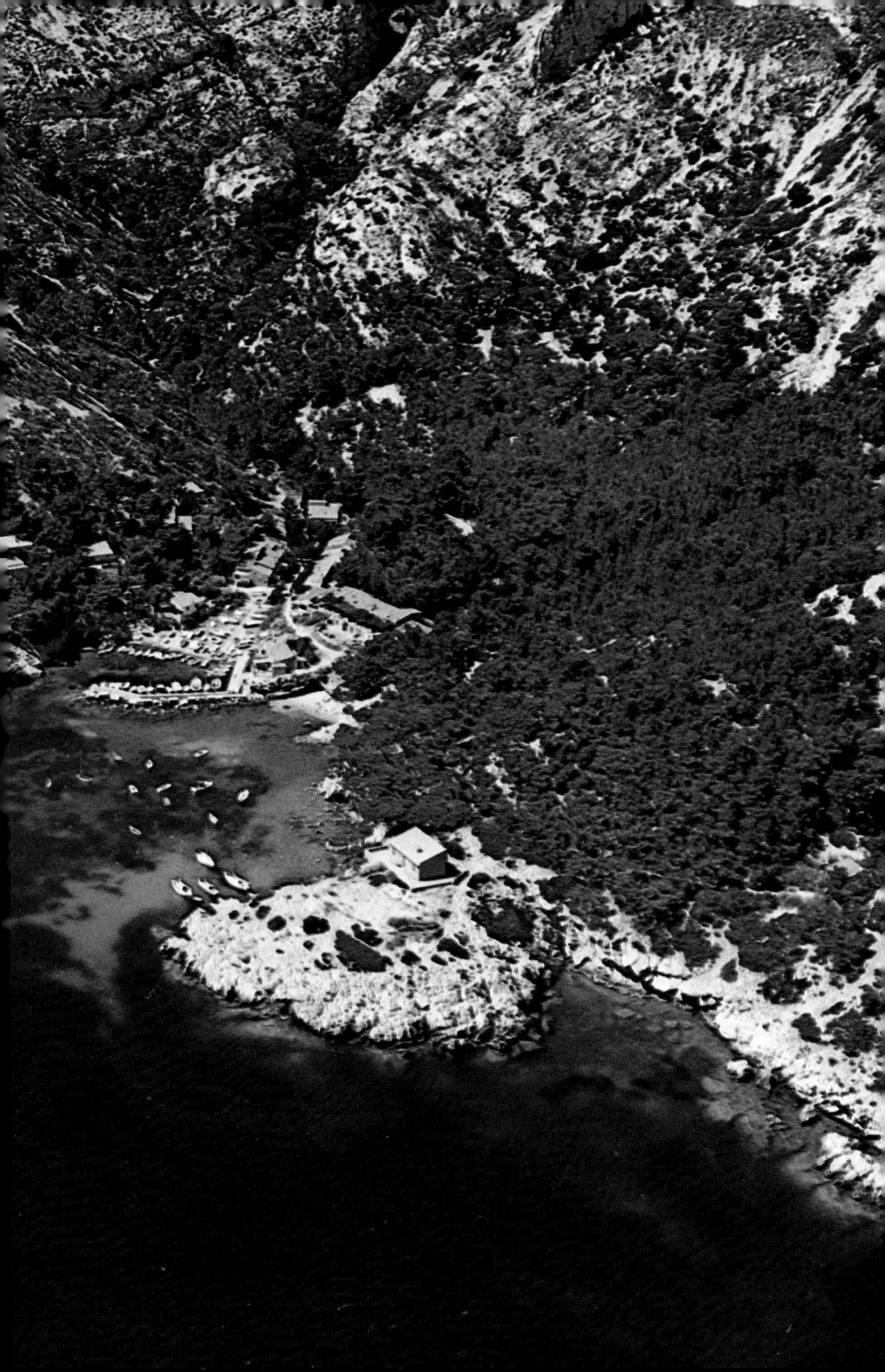

The four friends headed for Méjean through a tunnel a mile and a half long, carrying the grouper on a large branch. When they exited from the tunnel they were beneath camouflage nets covering German artillery. It was the grouper that saved them when they were caught. Backed against a wall by guards armed with machine guns, Falco and the others were being interrogated by an officer when he suddenly stopped to admire the size of their fish. He asked how they caught it, got a demonstration, and ended up letting them go. They ran along the railroad tracks as far as the little station of Niolon, where the stationmaster let them weigh the grouper on freight scales. It came to over 50 pounds.

Negotiating the rain and wind

He found a job. He had dropped out of school early and since 15 had been working in a company where his father had been a manager. But office work did not really suit him and the three years he spent there filing and handling bills seemed like a prison term. From noon until two o'clock he managed to escape long enough to take walks along the shore. Saturdays he and his friends took off on their bicycles as far as Berre Cove to swim and fish.

"That's when I learned to read the wind and currents, the very life of the sea," he says, "to negotiate rain and wind."

The group of friends, which now had a small fishing boat, discovered a new game at Sormiou. It consisted of taking the boat out in the worst weather and putting it through its paces. This gave Falco training that later would save his life.

"In a little bark that they call a *plate* or a *bette* in Marseilles I went as far as the Isle of Levant, even when the *mistral* was blowing hard," he says. "During some storms I was sailing right in the middle of submarines and warships. I kept my billfold in my bathing suit, and I figured that if I fell overboard I would be able to hold out for several hours." But the sea generously repays such courage and effort.

"The Isle of Levant is a paradise," says Falco. "I would dive three times and come up with three big grouper. I did it not only for the fun of it, but because we needed the money. We were very poor and what I got from selling fish was an important source of income for us."

Mine clearance

Falco was 18 when Provence was liberated. He and his friends felt particularly happy, for once again they had free access to the sea. Sormiou Cove had been dynamited and the fish were now less plentiful. The Germans had left mines and grenades around the boulders and buried in the sand of the beaches. They had thrown rifles, pistols, and ammunition into the water. This meant the beginning of another sort of game for Falco and his friends. Without giving a second thought to the danger involved, Falco volunteered to help the professional bomb squads clear out the area. And being an excellent swimmer and knowing the area well, his help proved invaluable. The main job was to clear out the large mines that had been placed on concrete pyramids along the beach. The group also pulled about a hundred pistols and rifles out of the cove, put them in crates, and took them out to sea for dumping.

"One day," Falco recalls, "it had been raining. The rain had uncovered something in the sand that looked like a fountain pen. I'd never seen anything like it, and yet I suspected it might be some kind of explosive. I picked it up to take it to a crate. Fortunately I reached for it with my left hand because when I touched it I set it off. It blew up and tore off four fingers."

But although this accident obviously put Falco out of action for several months, it did not end his swimming days. As soon as he was back on his feet he took up swimming as much as he could. He found he could now stay in the water four or five hours at a time. "When I look back on my life," he says, "it seems to be tinted blue. The sea has been everything to me."

Trips

In 1950 the group used an old 15-footer that belonged to a friend, Etienne Paul, who lent it to them on the condition that they put it back into shape. They managed that task in one week and, naming it the *Surcouf,* they took off in it for Corsica. Besides Falco, there was Paul Brémont, Bob Prigent, and Henri Plé. The trip was both hazardous and splendid. Other trips followed, for the friends bought an old 30-foot hulk and took it to Sormiou to outfit it.

"It was a terrific job," Falco remembers. "The hull was enormous and terribly heavy. We were unable to get it into drydock the first Sunday we had it, and

Albert Falco at Loude Rock in Sormiou Cove.

Right: Flash photos taken at Loude Rock: a wrasse, two groupers hidden under a rock, squid eggs amid red coral, a Bryozoa commonly known as Neptune's lace, and a *Cerianthus*.

during the week the wind came up and pushed it half way across the beach. We had to work a whole night in the rain, with me lugging a three-ton pulley block up the side of a hill. We worked on that thing all winter, sawing the ribs, putting in a diesel engine, rebuilding the cabin. Finally, by early summer, we were able to sail for Corsica on the *Hou Hop,* as we called it.''

The first good scuba

While Falco was learning navigation this way, he was also getting to know the sea bottom better with a new device: the Cousteau-Gagnan scuba outfit, or Aqua-Lung.®

The first time he had tried a scuba like that with Beuchat it had left him disappointed. Besides, his ability to get down to 30 or 35 feet simply by blocking his respiration gave him the impression that he did not really need one. But one of his friends had bought a Cousteau-Gagnan and hesitated to use it. So Falco tried it in Cancéou Cove. He went down to 60 feet.

``The bottom was covered with lobsters,'' he recalls. ``I was amazed that I could move up and down with such ease, as if I were suspended in the water. That was when I discovered that scuba diving could be miraculous. I taught myself how to use it without any help, and in the process I made a few mistakes. Since I was used to blocking my breath, I sometimes surfaced with my lungs still full of air. One day when I did that I felt an awful pain in my chest. At the time I didn't know what a risk I was taking because no one had warned me. I had kept my earlier diving reflexes and the mistake I committed that day with a scuba could have cost me my life. *

``It was at about this time that I met Armand Davso. He had made a scuba with a smooth-working pressure valve incorporating a large membrane. He was older and more experienced than I was and he gave me a lot of helpful tips. But with his scuba I couldn't dive to more than 60 to 90 feet. I could already get down that far without a scuba and even without flippers, lasting at least two minutes without breathing.''

---

*Air locked in the lungs remains at the pressure existing at the depth at which it is inhaled. If retained until the driver reaches the surface, it expands in the capillaries and can cause fatal injuries.

Thus the first 20 years of Albert Falco's life prepared him to become an exceptional man in and on the water, able to row or swim indefinitely and to dive deeply.

Double page following: Figuiére Point at Sormiou Cove, 60 feet down. The walls of a small grotto are covered with fixed animals, notably ascidia and red coral.

A diver puts an amphora into a sling at the Grand Congloué site.

2

the wreck at Grand Congloué

PIONEER DAYS—CHRISTIANINI'S SECRET—CAMPANIAN POTTERY—UNDERWATER WORKSITE—FIRST DEEP DIVE—SEAWORTHY—THE VACUUM CLEANER—APPRENTICESHIP—FIRST FRIENDS—TRAGEDY—ALL-WEATHER OPERATIONS—A DIABOLICAL DEVICE—DIGGING TECHNIQUES—MODERN SHIPWRECKS

"While I was working in that office in Marseilles," Falco says, "I felt as if I were trapped, cornered, sentenced to death. At noon I would go over to the Old Port to see the ships, like a prisoner looking at the sky.

"Things were happening at sea. Everybody my age was talking about it. We had heard that a group of men were diving, exploring, and making discoveries. I knew how to dive too; I had even tried out some scuba gear, although it was amateurish.

"I had the impression that I'd never be around those people who lived in the sea, never go out with them to dive. I was condemned to work all my life in that office where I was suffocating. They said that Cousteau had left the French navy to spend all his time in underwater research. He even had his own boat now. All that seemed like a fairy tale to me, beyond my reach. All I would ever be was a Sunday diver, an amateur sailor.

"And then one day the miracle happened. My entire life was changed all at once. The dream became reality. Cousteau and his friends, whom everyone talked

Albert Falco puts a basket of pottery on a truck. The pottery will go to the Borelly Museum in Marseilles.

Above left: A diver works with the suction tube to remove mud from amphorae and pottery.

Opposite page, left: A diver recovers two Campanian cups.

about so much, were accepting volunteers to work on an ancient shipwreck that they had found off Marseilles. Sunday divers were being accepted. I became part of the team. It was a miracle."

Pioneer days

August 16, 1952. There is a light wind as a small, 130-foot ship of 329 tons drops anchor in the narrows between the large island of Riou and the smaller one of Grand Congloué, about six miles from Marseilles. The ship is the *Calypso.* The rocky island of Grand Congloué measures less than 200 by 100 yards and rises 103 feet above the sea. It is mostly gray and black stone where a few wild plants grow and sea gulls circle. Separated from the coast by a narrow strip of blue water, it is exposed to the wind, battered by heavy rollers from the south and east. It has always been considered inaccessible by local fishermen.

On the quarterdeck of the *Calypso* three men are putting on their diving gear: one is Jacques-Yves Cousteau; the second, his oldest diving companion, Frédéric Dumas, who tried out the first scuba outfit designed by Cousteau and the engineer Emile Gagnan. The third diver is younger and stockier, built like a bull. His name is Albert Falco. He is 25, a new member of the team. Fascinated by the prestige and the bright future of these explorers of the deep, he works on the ocean bottom as a volunteer.

What are these men and that ship doing at Grand Congloué? They are living an adventure that would have been inconceivable a few years earlier: they are going through the wreck of a ship 120 feet down, a ship dating from the Third Century B.C.

It was in the Mediterranean, right after the war, that underwater exploration began, thanks to the Cousteau-Gagnan Aqua-Lung®. The French navy had created a diving unit at Toulon known as the Underwater Study and Research Group. But in 1952 the public is not very familiar with this rather mysterious activity known as underwater research. It remains the privilege of a few pioneers including naval officers and adventurers who bring back the first coral and amphorae.

This is the era of the first accidents. Until that time the only victims of underwater work had been professionals of the trade who wore heavy diving suits with helmets. Now a new race of men is born: free-floating divers who have not yet learned their potential and their limits. They have to undergo the apprenticeship of the deep. Some earn their living at the bottom of the sea thanks to the new scuba gear, and among them are those who believe that courage is all it takes to overcome any obstacle, even physiological danger.

It was one of those divers who dramatically revealed the wreck at Grand Congloué to the first underwater archaeologists.

Christianini's secret

A year before, a man had been brought into the Toulon naval base nearly paralyzed. He was taken to the Underwater Study and Research Group. It was obvious that he had made a dive and stayed down too long and too deep, and he had surfaced too fast. He was the victim of a decompression accident. This sort of accident was still relatively unknown at the time, but the man was immediately placed in the Group's pressure chamber and given first aid by a specialist in diving accidents. Several naval divers came around to see him during his convalescence, and it was to one of these, Frédéric Dumas, that he tells his story.

His name is Gaston Christianini. He is an underwater fisherman who concentrates on lobsters and a few curios like Roman pottery and old anchors found on the bottom. But he knows nothing of diving technique, not even enough to respect the necessary pauses for decompression when coming back up. That day in a cove the inevitable happened: Christianini comes back to the surface, but he is more dead than alive.

Quick action at the Group saves his life, but his toes remain numb and have to be amputated. Frédéric Dumas goes to see him in the hospital. Christianini is touched by this gesture and decides to talk.

"I'll never dive again. I'm going to tell you everything I know about the depths of the sea, my trade secrets."

For him, the main secret is how to find lobsters.

"You'll find plenty of them in the big cargo ship sunk near Maire Island. Also around an underwater arch at the foot of Grand Congloué, near the piles of old pots."

Campanian pottery

Frédéric Dumas immediately mentions Christianini's secret to Jacques-Yves Cousteau, who has asked the French navy for a leave of absence to devote himself full time to underwater exploration. Along with the leave he gets the *Calypso,* a former mine sweeper that is to be the only French oceanographic ship for over ten years.

Curious, Cousteau and Dumas go to Grand Congloué a few days later. They aren't so interested in lobsters as in those "old pots" Christianini had talked about.

Jacques-Yves Cousteau has recounted that first visit to a shipwreck that soon would become famous:

"Frédéric Dumas and I dive around the island. With us in the aluminum scow that serves as our diving workshop is Fernand Benoit, director of the Mu-

seum of Provence Antiques. After a long and uncomfortable dive that took me in vain down to 210 feet, I start back up following the rock face. At a depth of about 120 feet I find an immense pile of dark-colored debris. This mound of mud, pebbles, and pottery covered with accretions is an ancient shipwreck. It does not resemble those I have already explored in Mahdia, Antheor, or Maïre: this one is much larger. With growing excitement I pry between two amphorae and find a pile of finely wrought cups resembling chalices. I bring them up carefully, and Mr. Benoit declares immediately that they are Campanian vessels dating from the second or third century B.C.''

Underwater worksite

That same day Cousteau decides to devote several months' work aboard the *Calypso* to this find. At a depth of 120 feet, at the foot of a desert island, begins the first true underwater archeological digs.

''I will bring everything up,'' says Cousteau.

A superhuman task for which he will need much money, much time, and a great deal of equipment. And men.

Amateur divers and others with experience at underwater fishing come from all over the coast to volunteer help. In Marseilles one group includes old friends

On the Grand Congloué rock Ferdinand Lallemand sorts out fragments of pottery thrown up by the suction tube and places them in a basket.

Above, at left: Frédéric Dumas loosens an amphora from the mud.

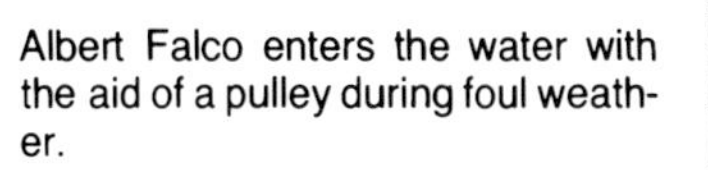

Albert Falco enters the water with the aid of a pulley during foul weather.

like Yves Girault, Armand Davso, and Georges Beuchat. On weekends, especially Sundays, the *Calypso*'s decks are crowded with as many as 35 persons. At times, seven divers are down at once. No one complains about discomfort or the food; the team is motivated by an enthusiasm that is already becoming known as "the *Calypso* spirit."

First deep dive

It was Georges Beuchat, in his small boat, who brought his friend Albert Falco to Grand Congloué one fine day and introduced him to Commander Cousteau, who had already heard of him from J. Borély, president of the French Federation of Studies and Underwater Sports. The very first day he set foot on the *Calypso*, Albert Falco made a deep dive. The suction pump that kept the digs free of mud had clogged, and Cousteau asked Falco to go down and unscrew the machine's head—at a depth of 127 feet.

"I went down alone," Falco remembers. "At the time I had some ear problems and had some difficulty keeping my balance. Still, I managed to go down, unscrew the big bronze nuts, remove the head and get it into a sling for hoisting aboard to be repaired. Commander Cousteau was happy with the job."

Falco returned several times to Grand Congloué as a volunteer and he handled a number of delicate operations. But he did have to earn a living after all, and had to keep on with his present job of ferrying and maintaining pleasure craft.

"Just about that time," he says, "I had to ferry a boat from Hyères to Marseilles. I still remember that black boat with big red sails, about 45 feet long. I spent a week around Porquerolles and Port Cros. Returning to Marseilles, I sailed as close as I could to Grand Congloué and saw the Commander and Madame Cousteau aboard the *Calypso*. They recognized me, waved, and the Commander shouted, 'Falco, I'd like to talk with you. See me Monday.'"

Seaworthy

"That was a Saturday. The *Calypso* was anchored in the Old Port at Marseilles on Monday, and Commander Cousteau said, 'Would you be interested in working with us full time?'

"I wanted that more than anything. It was the best break I ever had. I accepted immediately."

Commander Cousteau hired Falco for a trial period of one month. Then some administrative complications arose when the Maritime Commission objected. For Falco had one big handicap that could prevent his being hired permanently even as an ordinary seaman aboard the *Calypso*: he was missing four fingers on his left hand as a result of that mine clearing accident. He was very nearly declared unfit for the sea duty that he was so gifted for. But Cousteau knew what his new recruit could do. He himself accompanied Falco to his physical examination by the Merchant Marine doctor.

The doctor was reluctant to declare Falco "seaworthy." The regulations were clear: no one could go to sea missing four fingers. But Commander Cousteau insisted that Falco was just as able to hold a hawser as anyone else. Finally Falco was authorized to go to sea, but only aboard the two boats belonging to the French Oceanographic Cruises—the *Calypso* and the *Espadon*.

The vacuum cleaner

At the Grand Congloué site the divers used a suction tube that they called "the vacuum cleaner" on the mound of debris. It was a large tube that twisted, shook, and was very difficult to hold. It fully merited its other nickname, the "Loch Ness Monster."

Each diver made two, and sometimes three, dives per day of 15 minutes each, separated by at least three hours of rest so that their blood could eliminate the dissolved nitrogen produced by the dives. The end of the diving period was signalled by a shotgun fired on board the ship.

The excitement of the discoveries, the novelty of the work, and the importance of what was at stake kept everyone in a continual state of high tension. This was the first time in the world that digs had been effected at a depth of 140 feet in an attempt to salvage the treasure from a 2,000-year-old ship. It was an exhilarating task with no prospect of personal profit, and one some even believed impossible. But the very disinterested, generous character of the work was what made it so attractive. The team was young, enthusiastic, made up of young men who were excited about a brand new vocation, so new in fact that they were not sure whether it was a sport, was scientific research, or was a treasure hunt. The adventure at

On a rocky bottom a diver is about to grasp a broken amphora amid red gorgonia.

Grand Congloué aroused the interest and eventually the enthusiasm of people all over the coast around Marseilles.

Curiosity brought numerous visitors aboard the *Calypso*: officials, foreign scientists, and even the general who commanded the armed forces in the Marseilles area, who made a dive himself.

Changeable winds and seas, often difficult around Marseilles, complicated the team's job. The deafening noise of the compressor and winch increased the divers' fatigue. And above all this background noise came the frequent bangs of the shotgun.

Apprenticeship

Falco was all ears whenever he went to the *Calypso*'s mess room. He listened religiously as Dumas and Cousteau talked about diving. He made mental notes of all their tricks of the trade, everything that concerned this strange human activity of feeling one's way 120 feet down in a silent, viscous, heavy world where death was the penalty for a mistake. He was never formally taught or given advice, for this was a new area in which everyone had his own techniques and figured things out for himself. But the men who braved this dark underwater universe tried to communicate tacitly among themselves about the traps and dangers of the Beyond. Thus a wink by Dumas, a quiet laugh by Cousteau was worth more than a long discourse on how to handle oxygen bottles that got tangled up, a face mask that leaked or an amphora that refused to come unstuck from the bottom.

In all, it was a great diving school. It was marvelous training to work on such an ancient, complex wreck, to get everything out without breaking it, to fill baskets full of amphorae. You had to learn to handle scuba gear perfectly, to find a firm position on the bottom for maximum leverage. Archeological work like this quickly becomes an obsession. The excitement of the dig offered terrific temptations, for example to stay down just one more minute to get that last pot, to find out what is behind the piles of stone and debris. It can be stirring to dig on the sea bottom, to discover strange objects by feeling through mud with your bare hands. Often the objects seem to be glued together. But what joy to come up with an intact plate, complete with a hollow in the center for sauce, the whole thing covered with that splendid Campanian black varnish.

First friends

"At first there was no official captain aboard the *Calypso*," Falco says. "Cousteau himself generally ran his ship and Frédéric Dumas was chief diver. Now that I was working full time on the site I couldn't help feeling proud when the volunteers came aboard on weekends or vacations."

Two new divers came as volunteers: Jean-Pierre Servanti and Canoë Kientzy. They had been parachute commandos in the French navy, afterwards trying to fish for coral and sponge in Tunisia. Then they had heard of Commander

Cousteau and came looking for him. They asked to work on the site, and, thanks to their experience and ability, they integrated smoothly with the team.

The *Calypso* was having trouble with its mooring at Grand Congloué, despite the fact that the Marseilles Lighthouse Service had installed a buoy there attached to an anchor and a heavy chain on the bottom. After the ship had been positioned a month at the site, heavy swells had broken a shackle. The buoy floated out to sea, leaving the anchor on the bottom.

The first thing to do was to find that anchor because Cousteau and his team had no other, and finances were tight. The only way to find it was to follow the traces the chain had left in the mud on the bottom. Cousteau called on Servanti and Canoë to handle this task, since they were the most experienced of the new divers.

Servanti went down first. For a while his stream of air bubbles could be seen by crew members on the *Calypso.* He seemed to have become disoriented at one point, as sometimes happens during a deep dive, because he returned to the surface and dived again. Finally he came back to the *Calypso.* It was already late in the day and visibility was low. Servanti said that he had seen two traces left in the mud by the chain.

The search was put off until the next day. In those days deep dives were exceptional, and the dangers associated with them were not well understood.

Next day, Servanti dived in and swam off in what seemed to be the right direction. Falco accompanied him in a little dinghy and monitored his air bubbles. After ten minutes the large bubbles that indicate that a diver is moving were replaced by clouds of the small bubbles that mean he is remaining in the same spot.

Tragedy

Falco continued his watch on the surface. A slight wind came up, making small waves that kept him from seeing the bubbles as well. Nearby, the *Calypso* circled and waited. Cousteau and several divers were aboard a small barge moored to the buoy, which had been found.

Twenty minutes passed. Twenty-five. Servanti was 210 feet down, the depth at that spot being well established. Falco began to feel uneasy. Cousteau, 60 feet away, asked him if he still saw the bubbles. The last ones had just disappeared. Then there was nothing. The choppy water made it impossible to see below.

Albert Falco and Riquet Goiran reinforce the mooring of the derrick that supports the suction tube.

''Something's happened,'' Cousteau said, his voice tight. ''We've got to get him back up.''

And it was Falco he asked to go down.

''Instead of going straight down,'' Falco says, ''I headed for the spot in the water where I had seen the last bubbles. At first I got turned around. I had to surface. Once I was re-oriented, I dived again, doing zig-zags about 60 feet below the surface to cover as large an area as possible. I found a little diving buoy, one Servanti had attached to his belt.

''I started to descend on my first deep dive in the open sea. I tried to remember everything I had learned, all the advice I had heard: slow rhythm, counting my fingers, looking out for symptoms of light-headedness. Everything was becoming black. I had never been down so far. The air from my bottles tasted bad. I had to remember to land on the bottom feet first, never head first.

''At about 180 feet I had the impression I had plunged into an opaque night of liquid mud. It was an extremely unpleasant feeling, completely new to me. I experienced mounting anxiety, but I kept on descending, and when I got to about 30 feet from the bottom the water suddenly cleared. It was very surprising. I have experienced that many times since; daylight reflects on the bottom and illuminates the water.

"I saw the wire of the small buoy that was attached to a concrete 'dead man.' Servanti had done his job, but 30 feet away he lay motionless in the mud.

"At that point I had to begin using my reserve air supply. My dive had already been long, and the effort had left me winded. I didn't know what to do; I was in danger of being short of air. At 225 feet down you use a lot. Nevertheless, I went over to Servanti's body, took his mouthpiece, and put it between his teeth. He didn't react at all; he was completely inert. I undid his lead belt and tried to lift him. That was when I realized that I might end up the same way. I was out of air.

"I surfaced as fast as I could, without stopping along the way for the usual decompression. I didn't think that Servanti was dead. I thought maybe he had only passed out and that we would be able to save him.

"When I got to the surface, I screamed. They say it was an awful scream. The Commander had told two divers to get ready, Jackie Ertaud and Yves Girauld. They jumped in right away. I took them to the little buoy and showed them which way to go. They found Servanti's body and brought it back. We put it in a small decompression caisson and the *Calypso* made for Marseilles at full speed. But it was too late: Servanti was dead. Drowned. Probably he had swum too fast too deep to show what he could do. He had lost consciousness just at the moment when he was accomplishing his mission."

The anchor still had not been found. Cousteau himself decided to go look for it. He had to go down twice to get it into a sling so the *Calypso* could drop it back in the right place.

Servanti's death almost ended the Grand Congloué experiment. Commander Cousteau was deeply moved and felt like giving the whole thing up. For Falco, that terrible accident remains "a real nightmare."

Still, work went on. It was to last more than five years.

All-weather operations

The *Calypso* remained off Grand Congloué during July, August, and September 1952. By November the weather was downright bad, with very heavy waves. To insure that the mooring held, Falco and Canoë managed to get a chain around the pillar of a natural underwater arch. That was the best anchor possible.

The shed on the Grand Congloué rock that sheltered the materiel, and from which divers entered the water.

But it was still extremely difficult to release the hawser that attached the *Calypso* to the mooring buoy, so Commander Cousteau decided to keep one diving team on land. Getting a team of divers set up on the inhospitable rock that was Congloué was no small matter. The top of the rock, 60 feet above the water, was leveled off and a shack constructed there. It was so small that four persons could hardly move, but after hours of wind and rain, after returning from dives when everybody was shivering with cold, the new quarters were a luxury. A fixed installation on land would perhaps enable the team to mount an all-weather operation and to stay on the site all winter.

A boom 75 feet long, held by guy-wires, and a winch were used to position the suction equipment, which previously had to be hauled overboard from the *Calypso*.

The Grand Congloué rock was so narrow that it was necessary to construct a

platform at the foot of the boom to install a compressor for the suction gear. Beside it was placed a generator to furnish electricity for the shack.

A little team of divers took up residence: Jean Delmas, who was in charge, Riquet Goiran, Canoë Kientzy, Robert Picassou (a diver as well as an excellent cook), and Albert Falco. Volunteers came on weekends.

For six months the divers worked hard every day. They dug amphorae out of the mound of mud and piled them on the bottom. It was still impossible to tell the shape of the sunken ship. Often with their necks broken, the jars formed an incomprehensible mass from which the divers pulled out everything that stuck out. It seemed that they would never finish with this pile of pottery cemented by encrustations of barnacles and seaweed.

When the weather was good on weekends, the *Calypso* sailed for Marseilles. On the bottom, the jars were packed in special metal baskets and hauled aboard.

Sometimes the divers' booty for a single day amounted to 200, 300, or 400 amphorae, plus hundred of other bits of pottery. But the archeologists wanted to find even the smallest objects like coins, oil lamps, and cooking utensils which could be of great value. Therefore all the debris picked up by the suction tube was carefully sorted and conserved at Grand Congloué.

At the beginning of the work, the divers had met with great numbers of fish around the wreck. But gradually the fish deserted the site, due to the movement and noise of the suction tube. The octopi stayed, however, finding that the jars made ideal homes. Occasionally one of them would grab a diver by the arm and slide his tentacles all the way up to his chest or mask. It was a rather unpleasant surprise that the divers managed to turn into a joke. They would catch the octopus and put it in the suction tube. It would end up 120 feet above in the sorting basket and give the archeologist on duty quite a fright.

A diabolical device

Handling the suction tube created numerous problems, and as time went on it had to be improved. Great effort was required to move the tube around, and such exertion is always dangerous 120 feet down.

In addition, the compressor drew an enormous amount of air. If a diver inadvertently put his hand in front of the tube it was sucked into the tube and the machine had to be stopped immediately.

It was a brutal and temperamental instrument, seemingly endowed with a cunning and hostile nature of its own. It was a sort of monster whose caprices were unpredictable. The divers knew that if one of them were injured, the suction tube could become stuck to the wound and draw out all his blood.

Sometimes the thing had a fit, twisted, and tore itself from the arms that strained to hold it back and headed for the surface writhing like a mad serpent. Other times it dived back rapidly toward the bottom threatening to hit a diver. The brass components on the tube made up an enormous weight. When the tube thrashed around like that, it simply meant that its mouth was stopped up, and that happened regularly.

To unblock the tube it was necessary to winch it up into a vertical position and take it apart at the point where it appeared to be plugged. Then a 50-pound weight was dropped in from the top to clear it out. That happened about once a day and was exhausting work for the whole team.

Work continued all winter, despite the *mistral* and an east wind. The divers went down twice a day. Twice a day they climbed down the iron ladder that had been cemented to the rock and that is still there. If the waves were too heavy and it was impossible to dive in normally, they entered the water from the platform by being lowered on a pulley. One way or another they had to get down to the site. Surfacing was even more hazardous: it was neither easy nor pleasant to hold in place at the decompression levels amid waves that sometimes reached 25 to 30 feet down.

The dives were carefully calculated to avoid any danger of getting the bends from decompressing too fast. Ten minutes were allotted for decompressing gradually on the way up. There were never any cases of the bends during the whole time the site was worked.

If the Grand Congloué wreck caused deaths, it was only much much later; when Commander Cousteau decided to leave the digs, some divers explored the area clandestinely looking for pieces left behind on the bottom. As a result there were four deaths over the next few years.

"The wreck was a formless mass," Falco recalls, "surrounded by mud with some of the amphorae sticking out. Many of the jars were broken, making a sort of masonry heap mixed with pebbles. After we began diving daily we began to realize gradually that beneath the heap there was in fact the shape of a ship with a bow and stern and a wide middle. It wasn't very sharply outlined, but you could see a long shape that resembled a boat.

"It was only after many long days of work with the suction tube that we managed to cut through to the center of the boat and see that there was a layer of Roman amphorae, a layer of Greek ones, and between the two were piles of pottery. Bit by bit we got through the hull and then down to the keel.

"There were thousands of jars and thousands of other sorts of pottery aboard that ship. It was endless: nearly 10,000 dishes and 2,500 to 3,000 amphorae."

Digging techniques

"We had to dig the amphorae out of the mud carefully," says Falco. "We began by getting at one with the suction tube; we couldn't simply pull it out by the

Albert Falco exploring one of the numerous shipwrecks to be found on the bottom around Marseilles.

At left: Gorgonia, particularly what Falco calls "mimosa gorgonia," grow around the porthole of the wreck of a cargo ship.

handles because they would have broken off. We had to tap on the vase so water could filter through the mud. Each time we pulled one of those heavy vases out—like a tooth—we stirred up a big black cloud of mud.''

It was impossible to carry the vases straight to the basket that carried them to the surface. They could only be moved a few yards at a time. The divers wrapped their arms around them and proceeded toward the basket in a series of little hops, taking care not to bump into obstacles or get tangled up. It seemed funny to Falco to see them hopping around the site as if they were carrying babies in their arms.

''Sometimes,'' says Falco, ''I brought back as many as 36 vases during a 15- or 20-minute dive. As to the other pottery, the amount varied from day to day. At times we could pile 30, 40, or 50 cups into a basket, but other times we had to work with the suction tube for 20 minutes to clear out the mud. There were all sorts of ceramic articles: bowls, dishes for serving fish, perfume vials.''

The experience that Commander Cousteau had acquired working on the sunken ship at Mahdia, in Tunisia, where he had uncovered the ribs of the hull, prompted him to look for what remained of the structure of the wreck at Grand Congloué. Digging deeply into the mound, the divers found the keel, ribs, and lead sheets that had covered the hull. It was not a very large ship, being no more than 60 feet long, but it was very wide and probably very tall.

Modern shipwrecks

''For me,'' Falco says, ''Grand Congloué was a lesson: by digging in the mud you could bring back thousands of amphorae and other pieces of pottery and 2,000 years of history. The sea was rich, richer than I would ever have suspected. All you had to do was to look and search. I learned how to see. I discovered that at the foot of certain isolated reefs near dangerous Mediterranean coasts there lay other vases and antique anchors that were the remains of so many shipwrecks.

''I also learned what a modern wreck was, some almost intact, others gutted. All along the coast of Provence, at many different depths, lay ships sunk during the war. Some had been refloated or cut up where they lay to clear the channels, particularly at Marseilles around Estaque. But there were still plenty.

''It's impressive to visit a dead ship. It's like a great stage setting. You feel anxious as you head for it, for you know it's there somewhere but you don't see it yet. Sometimes you descend straight down toward it and it seems to take a long time to appear. Other times you have to search the bottom for it. It is nearly always enveloped in a sort of haze until the last moment, when that seems to clear and the hulk surges forth.

"In the Strait of Messina I visited two ships that had collided during the war. One of them was an armed cargo vessel whose captain had tried to run aground before sinking. He hadn't had enough time, and his ship sank in 180 feet of water. We found it by using a sounding device aboard the *Espadon,* and we marked the spot with a buoy. When I went down with scuba gear I couldn't see a thing. I reached the bottom at 195 feet. I began to look around, and managed to make out a black spot to the west. When I got to it I was up against the tops of the masts. The ship had rolled over on its port side, like an immense phantom lying there. I went over to the bow. It was an impressive sight: an iron wall 60 feet high, with the anchor in place in its locker. But everything looked disembodied, engulfed in haze.

"The ship was inhabited: when I got to the bridge, dozens of grouper came swimming out like an explosion of rockets and disappeared inside the hull through portholes. The mainmast still had its shrouds and ladders in place. A cloud of pomfrets were swimming around it. A rockfish hovered near each winch and bollard.

"On the afterdeck a single cannon was aimed at the sky, a black line standing out sharply in the blue water.

"The ghostly appearance of this ship-cadaver was accentuated by large nets hanging from the port side that formed curtains marked with white spots.

"After my experience at Grand Congloué, I would have liked to do other work on sunken ships whether ancient or modern. It's exciting work. But I had to stick to my assignments for the *Calypso.* I was one of the divers appointed to scientific cruises for biological and hydrological research. And as a matter of fact, I felt attracted by the study of marine life. Gathering and capturing sea animals for the scientists who came aboard the *Calypso* to study certain specimens was also exciting, and probably I was more suited to it than to archeology."

One of the sharks that the *Calypso*'s divers met in the Red Sea: *Carcharinus falciformis*.

3

confronting sharks: mastery of fear

FACE TO FACE—JEBEL ZEBAIR—FIRST OFFSHORE OIL PROSPECTING—VERY HARD ROCKS—VICTIM OF AN ALBACORE—THE LITTLE SPERM WHALE—REFLECTIONS ON SHARKS

January 7, 1954. At 9:30 P.M. in the Toulon shipyard, the *Calypso*'s hawsers are cast off. This is the sailing that Falco has waited for so long. This time he will be part of a long cruise: the Red Sea and the Indian Ocean.

Excitedly, he observes Commander Cousteau use radar to sail out of the port. It seems to him to be a magic instrument and he would like very much to know how to use it. He is on watch until midnight.

The sea has been whipped up by a strong *mistral.* The ship is rolling badly, the crew has to hold on to keep from being thrown from their bunks and it is hard to sleep.

The next day at 2 P.M., as the ship is coming out of the Strait of Bonifacio, a heavy sea from the southwest obliges the *Calypso* to seek shelter in the lee of some islands off Sardinia. But this transit through relative calm is shortlived, and soon the ship is making way slowly toward Messina, hit hard by waves against her side. Everyone on board has the impression that the *Calypso* is going to roll over under the force of the waves. But she rights herself and continues courageously on.

Falco has had a good watch, but he is exhausted by it and falls asleep immediately. When he is called for his next watch he finds Commander Cousteau watching the radar on the bridge. He takes the helm. The ship is one mile from the Strait of Messina. The radar image is extraordinarily sharp, and makes it considerably easier to negotiate the Strait.

January 10. The servo-motor for the helm goes out and the helm must be operated manually. The sea is heavier, making for hard work keeping on course.

About 10 that morning, five dolphins begin swimming forward of the ship. Leaning over the bow, the divers watch them play enviously. They stay ahead of the ship without any apparent effort.

January 11. Falco spends his 2 to 5 watch under an avalanche of hail and rain. The sea is still heavier, making the *Calypso* yaw wildly. Fortunately, the helm's servo-motor has been repaired. There are terrific waves hitting the stern. Crew members going to read the ship's log have to hold on tight, for violent waves are sweeping across the afterdeck. At last the ship arrives in Port Said, where it fills up on diesel fuel and water.

Falco is at the helm during the passage through the Suez Canal, when several big oil tankers rock the *Calypso* with their wakes.

Face to face

January 18. The midnight to 2 A.M. watch passes normally. By morning the *Calypso* is plying the Red Sea. The beacon on Friar Reef reflects on the intensely blue water. Falco is delighted.

At 10 A.M. the ship passes beside a rock protruding from a great depth. The *Calypso* moors itself to a big oval-shaped reef. Commander Cousteau has chosen a spot he knows well to stage the first encounter his new divers will have with sharks. The crew is bustling feverishly to prepare for this "premier."

Here is how Albert Falco evokes an event that became an important date in his life:

"I was anxious to get into that crystal water," he says. "I remember very well that we were all on the afterdeck. We were going in circles, and the sharks were going in circles too, but around the ship. I admit I felt uneasy. I had already seen sharks, mainly when I was spearfishing off Corsica; a big one had followed me around the Iles Sanguinaires. He was interested mainly in the fish I had attached to my belt. But we never had any problems with sharks in Corsica.

"Riquet and Canoë are designated the first team, while I get ready, somewhat nervously, with Jean Delmas.

"It wasn't ten minutes before the first team comes back up in a hurry. They had seen sharks close-up and panicked. I can see this is going to be fun. Still, I've got to go. I'm a little afraid, but I must absolutely give it a try.

"It's our turn. We enter the water. At first I'm dazzled by the crowd of multicolored fish, but I look around anyway to protect my rear: nothing in sight for the moment. We swim along the coral face about 60 feet down. Delmas is appointed to gather specimens, while I assure our security. But the first shark I see is tiny, about five feet long. He comes straight at us. At the last moment he turns and disappears into the blue. Pretty impressive anyway. Several others come along and run for the bottom after having feined an attack.

"From behind my face mask I can see the coral wall on one side and on the other the open sea with sharks in it, numerous sharks swimming round and round.

"Right away our first reaction is to put our backs against the wall and look for a shelter hole, as coral fish do. We are armed—if that's the word—with shark sticks. Dumas invented them in 1951, and now we are using them for the first time. It amounts to a stick with nails planted in one end. The idea is to push the beast away with it, hoping the nails will stick in its skin and make it turn. We are not yet entirely sure of the effectiveness of a device we have never before used.

"I remember very clearly how, backed against the coral wall, I didn't dare come out of the hole where I had found shelter. But we still had to move around and survey the reef a bit. After a while, we saw that these nice little animals came and went, turned in circles, and decided not to jump us. We had moved about 150 to 200 yards, turning around frequently to watch the sharks, which came to within seven to nine feet of us.

"I can still see the whole scene: Delmas kept on along that bumpy coral face. For the first time I see black coral, which looks brown and sticky and resembles a big gorgonia.

"I start to feel a bit reassured. I am relaxed enough to admire the fish that go in and out of the holes in the reef. A school of tuna goes by in the blue, followed by dangerous-looking barracudas. That time we thought we were in for it!

"At the time, barracudas had a very bad reputation. Today we know that barracudas generally leave divers alone. I have never seen a barracuda bite anything at all. Where we were there were hundreds, a whole school of them.

"Novice divers always think barracudas are going to jump them because they open their mouth, show their teeth, and pretend to bite; they always look angry. But if you charge them, they back away and then return obstinately.

To find out more about the movements of sharks in the Red Sea . . .

"After going about 200 yards along the coral face we surface amid large swells that are breaking over the reef. We are carried along by the force of the waves, and the launch following us is washed onto the outcroppings of coral. Our face masks are torn off, the launch overturns. It spills out all its equipment on the reef. We have the sharks on one side of the reef, and on the other we are splashing around trying to find our equipment. There is a moment of panic. But calm is soon restored. We bail out the launch. We get our matériel together, then head back to the ship, where everybody has a good laugh at our discomfiture.

"All that day I turned over in my mind that dive and our encounter with sharks. They weren't as bad as we had thought. Before experiencing them I was hesitant and worried. Now I'm impatient to see them again."

. . . Albert Falco is about to plant a distinctive marker in this one's dorsal fin.

It was at this time, in 1954, that Falco made his start as quartermaster. One fine day the Commander left him alone on the bridge during his first watch. He was not alone, however. Louis Malle, who was not much more experienced than he was, had been asked to assist him. Full responsibility for the *Calypso* had been turned over to them right there in the middle of the Red Sea.

Cousteau continued to educate his protégé. He taught him how to find the ship's position with a sextant. He spent many long hours at it; he began by having him determine their latitude by shooting the pole star. "I remember that he was very patient with me," Falco says, "and it's thanks to him that I learned astral navigation."

Seen from the *Calypso*, the shores of the Red Sea appeared neither hostile

Reaching his arm through the cage, a diver attracts sharks with bait.

nor frightening. In fact, they were beautiful, being of a reddish color with purplish mountains behind. It was an entirely mineral world, full of deep color and illuminated by a brilliant sun.

The Straits of Jubal were bordered by a beach of blond sand over a mile and a half long that went all the way to the foot of the mountains. The Sinai dominated the sea, high and chaotic, russet and black like overcooked bread.

Then came the string of coral islets where sea eagles made their nests. Finally there was the black mountain, Zebaïr: a block of lava where thousands of gannets nested, making white dots all over it.

A manta ray weighing over a ton passes ten feet above Falco and Frédéric Dumas.

Jebel Zebaïr

January 22. It is in the morning, as Falco is ending his watch, the *Calypso* reaches the Zebaïr Islands, which are set in a sea of deep blue. The ship shelters between two volcanos, one of which is fringed with a beautiful beach sparkling with foam. Hundreds of pink flamingos flock in a nearby lake.

Boats are put in the water and one team explores the south of the island without finding a habitable spot. Rocks underfoot make walking difficult. Finally the ship drops anchor near a small beach southwest of the island for exploratory diving.

Here is how Falco remembers it:

"Frédéric Dumas and I get our gear on, dive in and follow the point of the beach. We go down to 60 feet and I see a rocky area at about 105 feet. Frédéric Dumas dives for the bottom. I remain near the rock face. I feel uneasy because beneath the overcast sky the interior of the sea is dark.

"A small shark rises toward me on the left. This one is not turning away. I find a hole in the rock and slip in, facing the shark. It keeps on coming and forces me to back farther in. I decide suddenly to swim toward it, holding my shark stick in front. I touch it right in the face just as it turns—and it quickly flees. I have a quiet laugh behind my face mask.

"About 45 feet away, Frédérick Dumas is occupied with another small shark. He is keeping it in sight, revolving slowly, but he does not see two others rising from the bottom and aiming at his flippers. I charge them, waving my stick, and they back off. But instead of swimming farther away, they start to circle us rapidly, looking very interested in us. Frédéric Dumas and I exchange a look. These sharks are too numerous and too nervous. It must be time for them to eat. I don't intend to be on their menu. The best thing is to surface. Just at that instant I am surprised by the appearance of an enormous manta ray. Dumas hasn't seen it yet and I shout to warn him. We are side by side as the gigantic beast passes ten feet overhead, gliding majestically and casting a shadow over us. It must weigh over a ton.

"I'm beginning to think that sharks and a giant manta are rather a lot to go through in one dive, especially for someone like me who is just getting acquainted with the Red Sea. We finally get back to the surface. I'm eager to feel the bridge of the *Calypso* under me again."

First offshore oil prospecting

After a rough storm off Djibouti during which Falco for the first time sees the foredeck of the *Calypso* covered by waves, the ship stops at Djibouti, where it takes on two English specialists: they will work with the divers on petroleum research in the Persian Gulf.

January 31. On the way to Makalah Cousteau and the crew meet a school of dolphins like no one has ever seen. There are certainly over 3000 of them. They jump 15 to 18 feet out of the water and fall back on their sides, raising great splashes of water.

February 5. At 8:30 A.M. the *Calypso* reaches the Persian Gulf. She enters a sort of majestic fjord dominated by completely bare, yellowish cliffs. It is an overwhelming and implacable setting. The ship goes all the way to the end of the gulf until it stops in front of a little beach with a few fishermen's shacks on it. Some of the crew disembark on the beach, but night is falling. For that matter, the French sailors are rather badly received by the natives, and it is just as well to put off the visit until the next day to keep the divers from being completely stripped by them.

The next day, fishermen surround the *Calypso* in their boats and ask for water. Cousteau gives them some. Falco gives a chocolate bar to a child. He spits it out and throws it into the water; he doesn't know what it is. All they eat here is salted fish.

Very hard rocks

February 12. The *Caylpso* picked up a gravimeter in Toulon to use in its search for offshore oil. The crew is also to take numerous core samples from the bottom. Work begins at Dās Island. The first time it is sent down, the core cutter folds up like an accordion. The mechanics spend the afternoon repairing it. On the second try it works, and the *Calypso* moves two miles farther on to take another sample. This time the cutter brings up only fine sand.

February 13. After two tries at samples on a very hard bottom, the tubes come back up bent as if by a giant hand. It will be necessary to see and touch this bottom that is too hard for even the most advanced instruments. But the water is very cloudy and even milky. The Commander orders them to be careful, and Falco and Dumas plan to go down with the anti-shark cage.

The sharks are not the only animals to worry about. The divers discover that sea snakes live in the waters of the Persian Gulf. It is very unpleasant to encounter them, even though they are not aggressive. But the idea that they are there is enough to spoil the pleasure of the dives. There is no protection against them, for they can easily pass through the bars of the anti-shark cages.

Once the cage is on the bottom, Dumas gets out. Falco follows him and the two of them break off bits of rock with a hammer and chisel. It is terribly hard, but the geologists still are not satisfied. For them the samples are not sufficient.

February 14. While the mechanics are spending the morning repairing the *Calypso*'s winch, Falco prepares the anti-shark cage for another dive. He attaches a basket to the bars that will contain all the matériel they need and in which they will place the rock samples.

The sky has been overcast and it has even rained since the ship entered the Persian Gulf. They say it rains in this part of the world only once every 30 years; obviously it is not the right year for a cruise there. Finally the sun comes out and the sea becomes calm. Is it really the good weather that one associates with these latitudes?

There is another dive to the bottom with Dumas. This time they take the pneumatic hammer. It is all they can do to knock off a few fragments of rock in a half-hour. This is not the ideal machine. The divers can do just as well with a simple hammer. The water is still troubled, and if the cage is not a completely safe shelter, it is at least a refuge. But it is still necessary to leave the cage and work without protection to set up the traditional device of oil prospectors, the gravimeter, which measures gravity in several given spots and thus detects the presence of oil.

It is a heavy and cumbersome device that has to be positioned with great accuracy. Often three or four divers have to wrestle with it in the open sea, fully absorbed by a difficult task that requires great concentration. During those moments sharks would have an easy time if they caught the men by their feet, but in general they stay away.

The *Calypso* lines up its stations on parallel lines five miles apart and 70 miles long. But the bad weather returns and it is no longer possible to moor the ship aft. Then there is a fog of fine sand blown over from the coast. Visibility is very bad. Everything must be closed on the *Calypso,* for this sand gets into motors, the kitchen and even the bunks.

The weather improves on March 9, and the crew is able to begin a new line of stations toward the east.

March 12. Falco and his comrades discover thousands of pearl-bearing oysters on the bottom. It is their first interesting dive in a month.

In the evening, all the talk on the foredeck is about the pearls. Using scuba gear it would be easy to scoop up lots more in a matter of minutes. Everyone would be rich. Unfortunately, this has already been ruled out. The Sheik of Abu Dhabi has asked Commander Cousteau not to touch these oysters which, before the discovery of oil, constituted the emirate's wealth. Cousteau has given his word on behalf of the whole crew, and has issued strict orders. But it is a great temptation to glide over this seemingly inexhaustible supply of oysters.

Albert Falco, Frédéric Dumas, and Jacques-Yves Cousteau aboard the *Calypso* in the Red Sea during filming of *World of Silence*.

Dumas and Puig in the shark cage as it is lowered toward the water.

Albert Falco taking a bearing with the sextant aboard the *Calypso.*

Lowering the gravimeter into the Persian Gulf during the search for oil.

"I have never seen so many pearl-bearing oysters in my life," says Falco. "We had picked up a few of them anyway that we hid in our swimsuits or clothing. But we never found anything but small, worthless pearls. They were very deformed.

"We did all sorts of acrobatics to get those pearls, opening them in the gunwales when Commander Cousteau had his back turned. Dumas had as much fun at that as any of us. Sometimes we played tricks like putting a fish eye in an oyster and giving it to a buddy. We joked so much about those pearls that after a while no one believed in them anymore. That was when Albert Raud, a Breton, cried:

" 'Look at the pearl I've got!'

"It was in fact a magnificent pear-shaped pearl. Everyone mocked him, saying, 'It's not a real one, it's too big.'

"He finally got mad, and Captain Saout heard us. He came and confiscated all the pearls, including the biggest. He kept it, and when Raud got married the Commander had it mounted in a ring and gave it to the fiancée."

This petroleum research mission, the first offshore prospecting ever done, required the divers to go down every day, not only to handle the gravimetrics but also to take core samples or for dredging. The specialists made up little bags of samples that they sent to London for British Petroleum. These little bags that held such promise for the future seemed very precious and mysterious. Later developments showed that this research had not been in vain. It led to the exploitation of enormous wealth, but those who filled the little bags got none of it.

It was in the Persian Gulf that Falco begins to play the role of sailor as well as diver. He loves to work on the bridge: learning to use the sextant, use radar, and everything else associated with running a ship.

His life is gradually assuming its logical destiny. The child who rowed for hours and took a little boat as far as Isle of Levant is now responsible for a ship engaged in a difficult cruise. The young man who liked to spearfish in Sormiou Cove is now the diver who gathers rock samples, despite sharks and sea snakes, and works with geologists. And he has become an excellent quartermaster.

"I got to know the *Calypso* and its potential as if it were a living being," he says. "It became a real pleasure to take the helm, to stay on course, to watch the radar screen. Until then my only navigation instrument had been an old compass. Now I was using modern navigation instruments. It was a very heady experience. In those days there was plenty of opportunity on the *Calypso* for a man who wanted to work and learn. Captain Saout taught me the rudiments, but it was Commander Cousteau himself who taught me to use the sextant and radar."

Work is going ahead. The crew has lined up its 200th station.

Victim of an albacore

Excerpt from Falco's diary:

March 16. "It's been a bad day for me. At about 4:30 in the afternoon we are making for Dās Island to calibrate a gravimeter moored to a buoy that we had positioned when we were beginning this job. I take the opportunity to give a diving lesson to Maurice Léandri, who has improved considerably. Unfortunately, my eye is caught by a school of albacore circling us. I can't resist the temptation to go back and get a spear gun. I hit an albacore weighing about 25 pounds, but it fights so hard that it manages to shake loose from the spear and head for the bottom. I forget the spear gun and go down to join Maurice, who has remained on the bottom with the kentledge. I look around and see my albacore with its head in the sand, completely disoriented. I unsheath my knife and stab twice at the head. Unable to penetrate deeply, I take the knife with both hands and stick the blade into the head with all my strength. But my left hand slides past the guard and my little finger, the last of the five still remaining, is cut to the bone at the first knuckle. I surface with the fish in my right hand and call to the doctor to get his surgical instruments ready. He gives me local anesthetic, and in the heat I feel about to faint. The tendon is in bad shape; that left hand of mine is really unlucky. Still, I get off with a few stitches and a big bandage. The worst thing is that I won't be able to dive for some time. In the evening I have a high fever and the doctor puts me in the Commander's cabin where it's quieter. The next morning I get up as usual to stand my watch. My morale is still good."

March 18 and 19. "Bad days. I wander through the companionways with my left arm in a sling. Fortunately no one is diving."

March 20. "We are near Dās, and the doctor happens to replace my bandage just as we are at the same mooring where I cut myself. The drainage tubes are removed and there is no infection. The doctor asks me to try to move the end of my finger, but I can't. The tendon is completely cut. I start diving again anyway."

April 16. Between the islands of Dās and Divina, the divers set up the last three stations to make a round 400. It's Falco who finds the exact position for the last two with radar. He also makes the last dive. Then back to Bahrain.

From then on the *Calypso* heads south. These are the last few days that land will be in sight. Falco has the sensation of leaving for the end of the world.

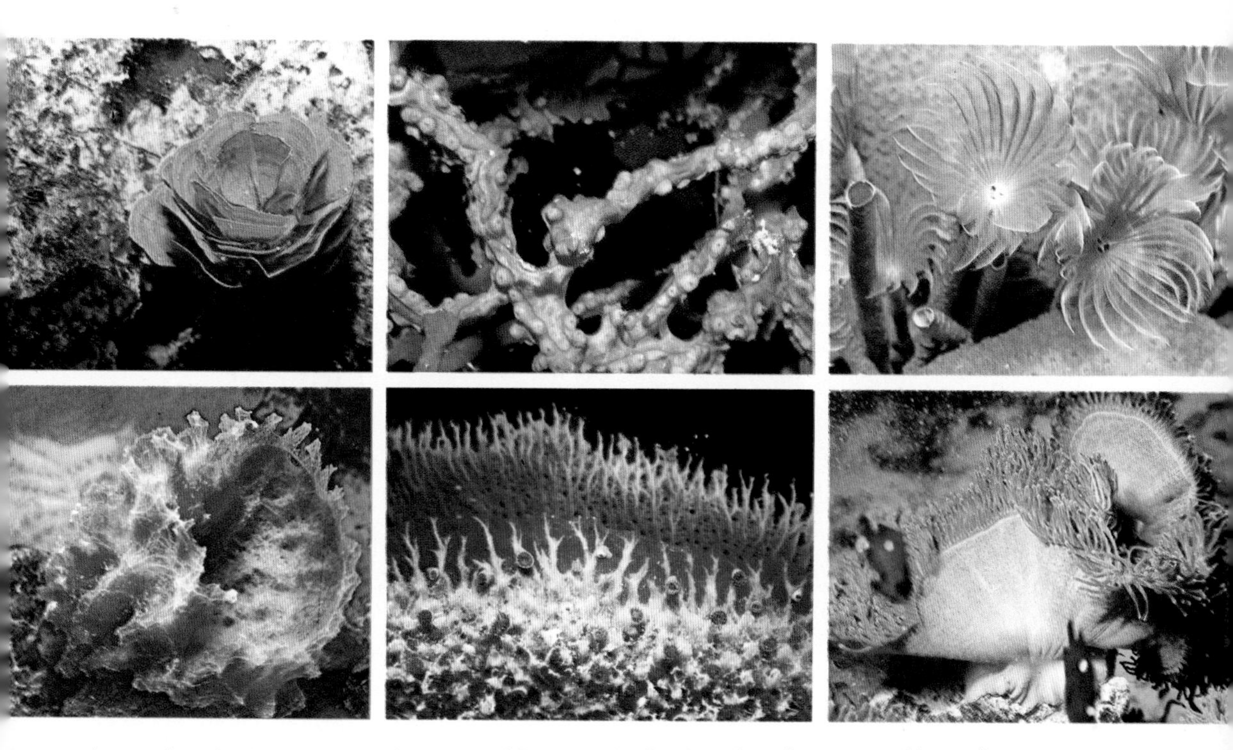

Images from the depths: above left: algae; middle: red and yellow branches of a sponge; a blossoming *Spirographis*, and a large sea anemone.

The crew of the *Calypso* is beginning one of those long voyages that it was to make many times through the Red Sea and Indian Ocean. But Commander Cousteau manages to stop the ship every morning near an island or a point of land so the divers can go down. There is no lack of surprises.

The little sperm whale

March 28. At about 8 A.M. a group of sperm whales can be seen not far off. The Commander starts after them with the *Calypso.* The whales, seemingly no more than large black masses in the water, maintain their distance easily. But all at once there is a bump under the hull and one of the engines slows down. A little whale has been sliced across the back five times by a propeller. It is losing a lot of blood, tinting the sea red with it. Its parents seem to crowd around. They surround it and give it support, but visibly it is losing its strength. Dumas puts it out of its misery with a rifle shot in the head. But the blood attracts a shark, then two, then ten of them, and after a half hour there are more than twenty sharks around the dead young sperm whale. Suddenly they begin biting off chunks of six or eight pounds of flesh from it. Their bites leave a hole the size of a dinner plate and three to four inches deep.

Falco goes down in the shark cage to film this terrible scene. It is a spectacle at once atrocious and fascinating. The whale's skin being very tough, the sharks seem to saw it with their teeth. They shake and twist their whole body to tear out a piece of meat. Falco shoots several rolls of film. The water is red with blood.

This sanguinary madness of the sharks has gotten on everyone's nerves. The crew feels a strange hatred toward them, and they try to get at them from the ship. Using gaffs, harpoons, lines, and rifles, the crew begins to attack them. A sort of orgy of blood ensues when other sharks arrive and throw themselves on the injured ones. The sea becomes a blood bath in which entrails, shreds of flesh, and skin float. Gradually, both men and sharks calm down. But Falco would like to see beneath the surface again to find out how the beasts are acting now. He goes down with Louis Malle in the cage. The sharks appear sated. They are swimming much more slowly. A big one, ten feet long as least, is immobilized, alone, about 30 feet from the cage. Falco opens the door, slips out and heads toward it. Malle follows with the camera. The shark stares at the diver with its cold eye. It doesn't budge. Six feet from it Falco stops. He no longer hears the camera motor. He turns and sees Malle signal to him from the cage: the camera is jammed. A quick look back at the shark: it is starting to swim slowly toward the man. Falco races for the cage, and Malle closes the door behind him. The shark keeps on coming and bumps against the steel bars. It seems disappointed. They take the cage back up.

During the night the *Calypso* ups anchor and leaves these depths that have produced such an eventful day. She makes for the Seychelles.

Coming from the Red Sea and the Persian Gulf, the Seychelles archipelago looks like a little paradise. Its first island comes up on the horizon like a green ball. After the black and russet burned rocks of Arabia, the shade of palm groves refreshes the eyes and the body. Falco discovers tropical islands for the first time, with their long beaches of white sand between the blue of the sea and the green of the palms.

The *Calypso* anchors at one island after the other so her divers can film the giant turtles of Aldabra. The lagoon of Aldabra is vast. It empties and fills with each tide. Cousteau and Falco enter the water at sea, at the mouth of the channel, with their camera, and let themselves be pulled along by the current, filming barracudas and sharks drifting along with them.

This cruise ends at Assumption Island, where the coral is more brilliant than in the Seychelles, and where Falco makes up to six dives per day. There the *Calypso* begins the trip back, meeting more sharks on the way back through the Red Sea. But now the divers have more self-assurance and they they wield their shark sticks with authority.

Reflections on sharks

Since that time Falco has acquired considerable experience with sharks. This is what he has to say about them:

"My years of apprenticeship spearfishing around Marseilles taught me many things about fish that have been useful in confronting sharks. Up to a point they behave like most other fish. You need to have swum among them in different parts of the world to know how to approach them and to evaluate accurately how much danger they represent. After a while, you begin to understand them. But never completely, for they always have some unpredictable reactions. When you are in the water with a shark you can sense very well whether it is passive or is interested in you. It is quite evident.

"Contrary to what you might think and contrary to most myths, a shark does not go for a diver right away. Of course, there have been some sharks that swam up to me quickly, but those are small ones, never big ones. They come charging up, but always stop a few yards off. They come and go, they turn in circles. They try to figure out what the diver is. Three-fourths of the time they go away. That leaves the last fourth.

"When a shark wants you, it doesn't go away, and if you are alone in the water without a cage and with no rock face to put your back against, it will get closer and start making smaller and smaller circles.

"Sharks behave the same way if you bait a line and drop it in. They never come up directly to it unless they have already been baited with a few bits of meat. The first time they see the bait they circle it once, twice, three times and then come up and bump it with their nose. Finally they make up their minds and bite.

A warm-sea shark with a coral backdrop: *Carcharinus albimarginatus.*

Above, right: Passing pompanos make silvery flashes.

Below, right: A Serranidae in the Red Sea spotted with blue points, *Cephalopholis.*

"There's no doubt that it is easier for a shark to bite bait than to bite a man, which he distrusts and for which it is an unknown creature. Luckily for us! For if sharks attacked men as soon as they saw them, no diver would ever put a foot in the water.

"From the moment a shark has taken its first bite, a sort of fever seems to take hold of it. It starts when the shark approaches closer and closer to its prey, and thanks to experience a diver can tell exactly when the animal's aggressiveness will reach the danger point.

"The disproportion between the strength of the man and that of the shark is considerable. The shark is in its element and does as it likes. Of course you can try to shoot a shark with an underwater gun, but even then you have to hit it in the right place.

"While operating a diving saucer along rock faces at great depths in the Red Sea I saw sharks six to nine feet long charge small fish or shrimp at incredible speed with their jaws open. Their speed is terrifying. You get the impression that a shark becomes angry when it misses a fish or even a shrimp. It is only at the instant it attacks that you can appreciate the power the animal has, for most of the time sharks swim slowly. You almost never see them at full speed.

"In the night of the depths, I have seen that terrifying spectacle many times in the light of the saucer's floodlights: a shark making its charge. It needs to be emphasized that the shark is not only a bundle of muscles, it is also an unbelievable nervous system that mobilizes the whole body. That is the reason for the shark's fury. When this nerve-driven machine starts moving, nothing can stop it. That is why a shark that has had its belly ripped open and has lost its viscera will keep on biting and swallow great mouthfuls. I have seen that happen.

"Most of the time a shark swims calmly and slowly. Its high-strung strength really appears only at the moment when its whole body shimmies as it plants its teeth in flesh and it goes through violent contorsions from head to tail to tear off a chunk.

"There are sharks and sharks. The small ones that remain mainly in the shallows and find all the food they want are the least dangerous. They rarely hunt and catch fish. Then there are those that stay in the depths around reefs, where there are plenty of smaller fish to prey on, and lastly there are the sharks of the open sea. Those can go for weeks, maybe months, without eating. Not having swallowed a fish for several weeks, they are ready to devour anything.

"On the *Calypso* we were in the middle of the Indian Ocean where the depth was about 10,000 to 12,000 feet, and where there were very few fish. All at once we saw sharks everywhere.

"Sharks might also follow a school of tuna. They swim all around the school or even right through it without biting anything. But all it takes is for one of the tuna to be injured for the sharks to swarm over it.

"Often I have pushed at a big shark with my shark stick; in those cases it's not the shark that moves so much as the diver, who is pushed back as if he has shoved against a rock. Sometimes, though, a shark will flee as fast as it can. That's not so much because the points of the stick hurt it as because its nerve ends are so sensitive that it feels the stick as sharply as it would an electric current.

"Sharks are very sensitive to blows on the head because of the sensitivity of their nervous system at that point. As a matter of fact, it is the mere contact rather than the violence of the blow that creates the desired effect. The stick's points actually are not intended to hurt the shark, but simply to keep the stick from sliding off its skin too easily."

Falco's example shows that the apprenticeship of a man is never over. When dealing with animals he can always learn techniques that enable him to overcome brute force with intelligence—even at the bottom of the sea.

4

from the Aegean Sea to the Red Sea

WITH THE SPONGE FISHERS—IN THE DOLPHINS' WAKE—CAPTURE—SEA FOREST—NIGHT IN THE FOREST—NIGHT FISHING—WINDOW UNDER THE SEA—AIR MAIL FISH

As early as 1956 Albert Falco acquired considerable experience in diving and underwater conditions. All the opportunities he now has had on the *Calypso* with the Cousteau crew are added to what he learned during his first explorations, either alone or with Beuchat off Marseilles. Not only has he followed intensive training during many missions, his qualities as an observer have also enabled him to acquire an exceptional knowledge of marine life—to compare the animals, scenery, and different aspects of the sea, which he has encountered during his professional activities. All this has gone into the making of his unique personality.

The year 1956 was marked by a cruise in Greece that revealed to the *Calypso*'s divers very different conditions than they had known heretofore.

The waters off Marseilles where they usually dived are characterized by sheer cliffs that drop off rapidly to great depths. In Corsica, where Falco had thoroughly explored the sea, the drop-off is not as sudden, and the bottom is covered

A dolphin plays in the wake of the *Espadon*.

by fallen boulders often covered by seaweed, creating a varied setting for sea life. But in Corsica, as around Marseilles, as soon as divers reach 90, 120, or 150 feet they find the same warm colors in their lamps reflected by fixed animals: the mauve of gorgonia or the yellow of sponges. In both Corsica and Provence grouper, sars, sea perch, and mullet swim amidst the seaweed on the bottom.

"Greece seemed much less rich to us," Falco says. "Brown seaweed supported by little floaters full of air is much more abundant than in Provence. But it was probably in Greece that we saw the most limpid water in the Mediterranean.

"We dived at least once in the waters of nearly all the Greek islands. We carried out many missions there, to conduct both biological and hydrological studies. But in general marine life there is poor.

"On the other hand, archeological vestiges are abundant, so much so that it would be vain and ridiculous to try to survey them in the course of a normal dive.

"At Cape Matapan (now Cape Taínaron), we entered the water at the mouth of a small stream and descended a 20-degree slope where seaweed grew lavishly. As we felt our way along, we discovered that the bottom was literally full of pottery, especially amphorae."

With the sponge fishers

The depths around Greece were reputed for their wealth of sponges. It was extremely tempting for the crew of the *Calypso* to dive with the sponge fishers and to film scenes from their dangerous profession.

Greek divers gathered together in the little port of Saint Nicolo, in the north of Crete, at the beginning and end of the sponge fishing season. The diver who was the principal actor for one of our films was a veritable giant, well over six feet tall. On the bottom he moved ponderously while the other divers came and went around him like flies around an elephant.

Still, he was agile in the water, and more effective than one might have expected. He knew his business well. He held a little hook in his hand with which he tore off the large sponges that filled a net sack suspended around his neck.

You have to have a great deal of experience and a good eye to spot the grayish masses in the water that are sponges, which have a diameter of five to fifteen inches. Sponge fishing is done at 90 to 120 feet down, and sponges are found only about every 90 to 150 feet. The diver, leaving a trail of small bubbles behind,

On the *Calypso* during the first tests of the electronic flash unit in 1953. In front: Professor Harold Edgerton and Jacques Y. Cousteau. Behind: Frédéric Dumas, Louis Malle, André Laban, Albert Falco and Yves Girauld.

pulled his air tube and safety cord along after him. He was followed by a little boat, the *Balancelle.*

Nowadays younger sponge fishers use scuba gear and even, in case of accident, decompression caissons. But in 1956, when Falco dived in Crete, the sponge fishers used only a diving suit with helmet and did not bother much about how long they stayed down or to observe pauses for decompression when coming up. The skipper of the *Balancelle* knew a little about such things, notably that divers must not stay too long on the bottom or go too deep, for he had witnessed several accidents. Actually it was he who decided everything, particularly the length of the dives.

Those divers were in fact well trained. They did not use a ladder to enter the water, but stood on the gunwales forward and dived in. The boat followed the diver, motor idling, the skipper holding the safety line.

Sometimes there were accidents, as when the diver had to climb over a rock and slipped into deep water. There was no first-aid equipment aboard with which to treat injuries. But these Greeks knew of an empirical technique to treat the bends: they buried the diver in the sand. They said that they had managed to save several of their people like this.

In the dolphins' wake

Already initiated in marine biology, Falco now became interested in bigger sea animals like dolphins. His contact with them was both more difficult and more exciting.

He thus begins to discover his true vocation and the gift he has for approaching, reassuring, and almost taming animals from a different world—animals that have never been in contact with man.

Falco knew that even gigantic animals that seemed beyond the reach of human beings showed a behavior and an intelligence that he dreamed of understanding. He had learned this in Corsica around the sperm whales he had encountered there, as well as during his work on the *Calypso* in the Indian Ocean.

Now he had a new opportunity. It was to devote his time to working with those marine mammals that have always shown a mysterious sympathy for man: dolphins. In 1957, Commander Cousteau asked him to try to capture one or two for the Oceanographic Museum of Monaco.

"I put a lot of effort into learning to approach, touch, and even caress dolphins," says Falco. "My hope was that one day they would become friendly and accept my presence.

"The story of that first try in 1957 has already been told.* First it was necessary to invent a technique for capturing dolphins. I had put a plank out over the bow of the *Espadon* and tried to grab dolphins out of the water from there. As soon as one was caught we had to pick it up in a launch and hoist it aboard. That was the most delicate operation of the hunt, but often the dolphin was so traumatized that it did not struggle much. We brought it back to the museum as fast as we could and put it immediately into a pool. The ones we caught were usually the species *Delphinus delphis*.

"We could not leave it by itself for a while because it would either be unable to breathe and sink or would run into the edge of the pool and knock itself unconscious. I got into the water with it. I supported it with my right hand on its dorsal fin and my left hand on its ventral fin. I moved it slowly around the pool so it could familiarize itself. I got it to touch the edges lightly with its nose, and I dived with it to lead it down to the bottom. After 15 or 20 minutes the dolphin usually began to stay on the surface and give a few kicks with its tail. Then it would begin

---

*See *Les Dauphins et la Liberté*, Jacques-Yves Cousteau and Philippe Diolé, published by Flammarion, France.

The *Espadon* in port at Monaco in 1956.

You have to get close to dolphins in the water to understand them.

to swim slowly in circles. I knew then that the main danger was over. The dolphin had understood that it was a prisoner and it knew that it must not hit the pool's cement walls. Unfortunately, some of them would literally kill themselves by swimming at full speed into the walls. That is what I was trying to avoid at all cost."

Capture

"Never once did any of the dolphins try to bite me, and yet I know that several of them suffered from sickness and from the effects of being captured.

"We kept one female for six months. In those days, only the Americans had attempted such captures, and we didn't know much about dolphins. No one had yet managed to keep them in captivity and to be accepted by them.

"I have certainly kept many memories from that experience that have left their mark on me. I was struck by the impression that several of our prisoners had a real feeling for man and in particular for me. One female, for example, gave cries that could only be interpreted as cries of joy when she heard the door to her pool squeak open. It was an extraordinary feeling to think that our presence could console these animals that were suffering from the solitude of prison.

"All my experience with marine life, including my first dives at Sormiou, my contact with fish, and my work beneath the sea, certainly was useful in this first attempt at approaching marine mammals.

"With them I learned patience and above all what I would call a respect for their personalities. You could almost say that each dolphin is unique, that it is like no other. We got to know all sorts: funny, stubborn, desperate, melancholy and loving."

To acquire such convictions you have to have known these animals in their element. Anyone who has not done so probably sees that as silly anthropomorphism.

Sea forest

One of the rules of the Cousteau crew was that each member be able to fill several roles and to know many different techniques. Albert Falco soon had to renounce his role of animal trainer to resume his place aboard a *Calypso* about to begin a new hydrological research cruise in the Mediterranean. This time he was to be navigator. His captain was to be Commander Saout.

Falco's task was often delicate. To accomplish the mission set out by the National Scientific Research Council he often had to keep the ship in the same spot for four hours by constantly maneuvering.

In 1958 their research resulted in the chance discovery of an extraordinary submarine forest near Alborán Island, south of Málaga.

"It was during a *Calypso* mission," Falco recounts, "that we came across that marvelous marine vegetation. Professor J.M. Pérès, director of the Endoume marine research station, was aboard to do some biological research. While sounding at 120 feet we observed the presence near Alborán Island of an underwater hillock that seemed inexplicable. We dredged and came up with laminaria of considerable dimensions, up to 25 feet long.* The *Calypso* dropped anchor and I entered the water. I found an extraordinary spectacle: a veritable dome of vegetation made up of greenish-brown runners and leaves. When I swam over this forest, grouper came out of it by the dozens, like rabbits running from a wood. The current is very strong and makes the big laminaries wave and twist like serpents. But as soon as I dive into it, the current stops. Within the seaweed everything is calm, as in a forest where the trees cut off the wind. We are the first to penetrate this jungle.

"We can wander easily among the stems, which are attached to rock by strong roots that resemble the tentacles of an octopus. A group of albacore follows us in the forest. They come right up to us, file by, turn, and leave. Their perfect silhouettes are profiled against the light.

"Illuminated by my lamp, the rocks seem marbled with yellow and red. Hiding in the algae, a moray bobs in the current, eying its prey. We carefully avoid it. Several large pink fish rise from the bottom, a hydroid attached to the head of one. There are also lobsters and numerous mollusks. Marine life is extremely rich in this forest.

**Laminaria ochroleuca*. As of this date, this is the only place in the Mediterranean where algae this large have been found in such abundance.

Double page following: Falco attempting to calm a captured dolphin in the Palm Beach swimming pool in Monaco.

Albert Falco on the bridge at the controls of the *Calypso*.

"As in all the world's forests, it would be easy to get lost in this one. Fortunately, the water is clear and our mooring is visible from far off."

Night in the forest

Commander Cousteau had turned over a former trawler, the *Espadon*, to Falco for another mission, and he returned to Alborán shortly thereafter.

"The *Calypso*, which was nearby, came to join us," Falco recalls, "and we made some very good dives. To hold against the current we let down some ballast to the level of the algae and held on to it while filming. The morays are still looking for prey in the vegetation and stick their heads out suddenly when we pass by.

"A large group of us also made nighttime dives. But the current hampers us in lowering the floodlights and cameras. The Commander tries something else: a sort of submarine troika that we hang on to with our equipment. An astonishing spectacle unfolds before us. The current carries along luminous jellyfish and salpas that float past us and disappear into the night. Other jellyfish covered with brilliant multicolored dots and large white filaments with milky spots catch on the cable that the current sets vibrating in our hands. We reach the submarine forest without too much difficulty and set up for the filming.

"Under the floodlights, albacore throw off silvery sparks as they pass into the shadows. Everything seems to be going well, when suddenly one of the crew, Riquet Goiran, passes out on the bottom. Canoë notices it first. He lets go of his camera, which falls into the algae. He grabs Riquet, who has a moment of panic, pulls him over to the cable and hoists him up. The Commander, who was zigzagging among the stems for a traveling shot, comes out tangled up in a plant that nearly covers him. Then we head for the *Espadon* along the cable above the forest, going in and out among the plants like insects among flowers.

"It must be admitted that it was not the best subject for filming because all that can be seen is a continuous wall of plants. But however disappointing the work may have finally been, it left each of of us with the memory of a unique place, a little like something from Jules Verne."

Falco already knew something of shooting underwater films. He was beginning to learn how to handle the camera. He had learned how to set up the floodlights and to select the right angle for a shot. Still, he had never been a professional filmmaker. There had been many of them aboard the *Calypso,* including Louis Malle, Jacques Ertaud, Michel Deloire, Jacques Renoir, and Pierre Goupil. But at Alborán the professionals felt uneasy 120 feet down and most of the time Falco did the filming.

Night fishing

It was about this time that Falco did considerable night fishing, and also shot some documentary films for the Monaco Museum. When he was preparing to try out some underwater cameras at Sormiou Cove, he noticed that sea cucumbers were standing almost vertically on the bottom, and that they contorted themselves in a way that resembled snakes. They also threw off clouds of white matter that dissolved in the water. He had the camera with him and filmed this chance scene.

This nocturnal visit with the sea cucumbers also led him to discover that the fish caught by his floodlights were completely blinded and immobilized. After that, the Monaco Museum asked him to go night fishing several times, and he brought back over 150 fish. Sar, sea perch, and mullet are easy to take at night. A landing net is put in front of the fish and the lights are shone directly in its eyes. The net is brought slowly toward the fish, and when its head touches the thread it jumps straight in. The net is then folded in two. The technique is a bit like catching butterflies.

A diver above the Alborán algae forest.

A diver penetrates into the forest of large Laminaria.

Right: Raymond Coll and Albert Falco with one of the Laminaria on the deck of the *Calypso.* It measures 22 feet long.

This works for all species of fish that swim in the open water, but a different tactic is necessary for fish settled on rocks. In the case of the red mullet, for example, the net is put on the bottom a few inches from its nose. It will advance on its wattles and climb by itself into the pocket of the net. But capturing a grouper is real sport. It is rare to find one in the open water at night. When one is found, the net is put at the entrance to its hole. The fish is touched lightly and the lights are turned off for a fraction of a second. The grouper makes a jump and runs into the net with such violence that it sometimes tears it. The fish can be calmed by blinding it with the lights all the while it is being brought to the surface. That is the only way to bring it back intact.

Such night fishing has enabled the Oceanographic Institute to fill its aquariums for a long time to come.

Window under the sea

Falco's success at this sort of fishing led to an assignment farther away. He was asked to go to Elath (now Aqaba), on the Gulf of Aqaba in the Red Sea, to gather coral and tropical fish for the Monaco Museum. Georges Alépée and Pierre Goupil were to film a short documentary on the same trip entitled *Window Under the Sea*. This is Falco's account of that expedition:

"Israel gets very hot in the month of May. The plane that brings me and my little crew to Elath, an old DC-3, goes through one air pocket after the other. There are three of us: the filmmaker Goupil, Georges Alépée, and myself.

"It is stiflingly hot when we land. The tarmac is beginning to melt in places. It is Independence Day and no one is working, so we have to unload our equipment ourselves, and there is a lot of it, including a compressor. We start on foot for Elath, overwhelmed by the sun and perspiring heavily.

"When we get into the water we discover some very beautiful groups of coral of different colors and shapes, isolated in clumps. I feel optimistic again. The reefs of the Red Sea and the Indian Ocean where I have dived were perhaps more varied, but they were larger and offered almost too much choice. Here I will be able to work in greater detail, observing and filming the fish and their setting.

"But on the first evening Alépée and Goupil are terribly red from the sun. Alépée shivers with fever all night.

"I make my first dive the next day. I want to begin collecting my samples for

Monaco. I catch a pteropod in the hand-net that we call 'The Madwoman of Chaillot' because of its winglike fins. A little cofferfish and spotted moray are caught in a plexiglass trap.

"Alépée is so sunburned that he can't walk. As for myself, I get the unfortunate idea of shaking the coral to make the fish come out. I'll have to find a better way because I immediately feel thousands of pinpricks all over my body. All the urticating cells of the coral have hit me.

"That evening I am covered with blisters. I have a high fever and don't get any sleep that night.

"In the morning a sandstorm hits Elath. The bad weather lets us get some rest and take care of our injuries. The crew is not doing very well. I try to think of a way to catch those coral fish that run to hide in their holes.

"To get myself back in shape I make a deep, 45-minute dive. I spot a large black coral stump with several branches.

"We finally manage to rent a truck to transport all our equipment. It is time to get going on the most serious parts of our job. But the truck gets stuck in the sand. We practically have to pave our own road to get it to the beach near the Egyptian border. That is where I want to set up our base camp. We struggle with the truck, the equipment, and the sand beneath a blazing sun. At 2 in the afternoon it is 45 degrees centigrade (113 F.) in the shade. Anyone who steps barefoot on the ground quickly gets burned.

"I am finally going to try a trick I have thought of to catch coral fish. All it amounts to is wrapping a net around an isolated coral outcropping. The configuration of the reefs in the Gulf of Aqaba lend themselves to this technique. I dive with a net 60 feet long attached to a metal pole with which I spread the net around the coral. As I wait motionless, several pomfrets are caught in the webbing. I can hear their cries, a very odd sensation. Other fish—one blue, one green, a parrotfish—fall into the trap. I have to get to them before they injure themselves struggling to escape. I take them out and put them in plastic bags, being careful to leave a bubble of air inside, and tie them to the coral. They make a strange sight.

"I have also found an amusing way to capture triggerfish, which we call "Fernandels" because of their wide eyes and open mouth. I put my hand in their hole, I press down the spine on the dorsal fin and easily get them out. But the biggest and most beautiful I came across gave me the most trouble. I finally had to use a hammer and chisel on the coral to get my hand in to press the spine down. It seemed so surprised that it didn't even struggle.

"Three more hot days: 40 degrees centigrade (104 F.). The air is unbreathable. But we have to shoot the exterior locations for our film.

Fish off a coral reef: above left, an angelfish *(Pygoplithes dicanthus)*, at right a *Holacanthus imperator* amid coral.

At left: These scarlet fish are *Myripristis*.

Below: A checkered triggerfish.

At Elath (Aqaba) Albert Falco captures fish destined to be sent by plane to the Monaco Oceanographic Museum.

"Georges Alépée leaves for Tel Aviv to contact Air France about transporting our fish. The Oceanographic Museum of Elath agrees to keep them in its aquariums until we can send them to Monaco, but a few fish die every day.

"I continue to take fish with Goupil. But our nylon net is in tatters by now. I get another big green triggerfish with a crescent-shaped tail, some surgeonfish, a yellow and white cofferfish, a marvelous leaf-shaped fish and a little "Madwoman of Chaillot." While we are transferring them to their aquariums a terrible sandstorm comes up and after a few minutes Elath becomes invisible."

Air mail fish

"Alépée comes back from Tel Aviv. The fish are to be sent on June 3 at 3 P.M. We have a lot of work to do to get them packed. Specialists from the museum help and advise us, as do several local divers, who have become friendly with us.

"Here is how to proceed: we pour five liters of water into a nylon bag and put one large fish and a dozen small ones in it, then pump it with oxygen . We put one or two other bags around this and fill them with air to keep the water temperature at about 27 degrees centigrade (81 F.). I'm afraid that these coral fish with their sharp teeth and spines may tear the nylon. Each recipient is enclosed in a cardboard box padded with kapok.

"At 7:30 A.M. all our boxes are loaded aboard the plane at Elath for the trip to Tel Aviv, where they will be transferred to the Air France plane. They will be in Nice the same day at 4:30 P.M.—a speed record for fish.

"We take the next plane. In Tel Aviv there's a surprise awaiting us: the local Air France manager, Mr. Ghys, invites us to his home for a party he has prepared for us. I am astonished to learn that in 1944 he invented a diving suit that I tried at Sormiou with Beuchat and Cazals.

"Back in Monaco all that remains for us to do is to decorate the aquariums reserved for our fish with the coral we have brought back. But that doesn't look quite right, so Canoë and I dive in the bay of Villefranche to bring back some gorgonia. That is just what we need.

"This assignment taught me many things. Above all I learned through direct contact with coral fish how much their lives are in danger. They are extremely fragile. In gathering these specimens for the Oceanographic Museum we tried to do as little damage as possible in that world of coral where all life is closely linked.

"I know that since then the Gulf of Aqaba has suffered a great deal from pollution, and that a lot of coral has died. Even at the time when we were there to film *Window Under the Sea* we could sense that the life of these magnificent reefs was in danger."

5

the diving saucer

FIRST TRYOUTS AT PUERTO RICO—EXPLOSIONS—300 FEET DOWN—DEEP DIVES—SERVING RESEARCH—THE CANYON OF CASSIDAIGNE—RECH LACAZE DUTHIERS—AT MINUS 980 FEET—SWEPT AWAY BY THE CURRENT—THE TWO EELS—A NEW WORLD

The month of July 1959 is an important milestone in Falco's life. That is when he first comes into contact with a device that will transform his vision of the underwater world and considerably enrich his knowledge and experience.

For it is then that the first experiments with the diving saucer are made. The saucer is a small submarine of truly revolutionary conception. It is compact enough to be taken aboard the *Calypso;* indeed, the ship would eventually transport two of these miniature submarines.

Cousteau first had the idea of constructing the saucers in March 1955, when the *Calypso* was anchored in front of the cliffs of the Red Sea. He was taking a bath and toying distractedly with the shower head in the tub. It was then that he conceived the notion of a round underwater device that would be propulsed, not

After surfacing, the S.P. 350 diving saucer sends spurts of water into the air so it can be sighted by the *Calypso.*

by propellers, but by nozzles in the water. The saucer at first is called the *Tortue* because of its shape and in memory of Bushnell's *Turtle.* *

Cousteau asks Falco, "Would you be interested in being the first pilot of that craft?"

Falco naturally accepts the offer enthusiastically, immediately recognizing the importance of the new machine. Thanks to the diving saucer, the limits of underwater exploration will be extended much further. Man's penetration into the seas will become possible as deep as 1,000 feet, considerably enlarging his underwater dominion. Moreover, visits to the bottom will no longer present the same dangers. Passengers in the saucer will be able to make virtual underwater tours without fatigue, and even if they remain several hours at great depth they will not be obliged to make the usual decompression pauses because they will have been breathing air at normal atmospheric pressure. A science fiction dream is about to be realized.

Unfortunately, it takes over two years to finish the prototype diving saucer, due to shortages of funds and time. It is constructed from Cousteau's blueprints by the French Office for Underwater Research in Marseilles, with the help of the engineer Jean Mollard and André Laban.

Finally in July 1959, the miniature submarine is finished and Falco works excitedly with the engineers on the finishing touches. This first diving saucer is launched on July 21 in the port of Marseilles. Firemen are on hand in case of an accident.

The tests are encouraging: the machine is perfectly watertight. The controls seem to work. But it is obvious that there is still a lot to learn before being able to handle a capricious new machine like this one. It is still too light, despite 550 pounds of ballast. That is just as well, since it affords a large margin of safety in discharging ballast. A single incident mars the first test: the pump motor burns out.

A second test in the port of Frioul goes less well. The hatch leaks and the two passengers have to get out in a hurry.

On March 16, the *Calypso* weighs anchor with the saucer aboard. The oceanographic ship carries out several missions during the month of August while crossing the Atlantic, including hydrographic work at Meteor Bank with soundings and dredgings with the troika.

---

*See Appendix: "From Bushnell's turtle to the diving saucer."

Falco takes his watch regularly and rounds out his navigator's training. When he has a minute he works on the saucer, particularly on the exterior fastenings for the batteries. This is extremely important: miniature submarines with interior batteries have often killed their occupants.

First tryouts at Puerto Rico

After stops in Bermuda and New York—where she is welcomed like the *Normandie* by fireboats shooting geysers into the air—the *Calypso* goes up the Potomac for an enthusiastic greeting and then heads for Puerto Rico.

Rain and a heavy sea delay the saucer's open-sea testing. Falco dives with scuba gear to explore the corals and gorgonias on the bottom and generally to compare the underwater scenery of the Caribbean with that of the Red Sea and Indian Ocean. The sea never looks quite the same in different spots on the earth, even if tropical coral reefs do present certain analogies. Soon the weather improves and the sea becomes calm. The first tryouts of the saucer are to be off the western coast of Puerto Rico.

At 3:15 P.M. on October 9, 1959, Falco and Mollard close the hatch and start the first dive. The diver assigned to accompany them is one of the most experienced of the Cousteau crew: Henri Plé, whom, for no particular reason, everyone calls "uncle."

For extra safety a cable links the saucer to the surface. It is too soon to release it completely. Aboard the *Calypso*, the whole crew observes the operation. The experiment is deliberately staged in a relatively shallow cove.

The dive gets off to a bad start: the saucer assumes a strong list to port. Inside, the tape recorder falls on Falco's head. At 3:30 it touches bottom at 49 feet. Inside, the pressure climbs abnormally, with the barometer reaching 36 inches.

Plé looks in at the porthole frequently. Falco gives him the O.K. sign. He runs through tests with the nozzles, shifts the weights around to correct the list, tries to maneuver around the corals and gorgonia, and lands on a fairly flat bottom. Then he begins the ascent, trying to keep it slow. The dive has lasted 40 minutes. When the hatch is opened on the surface the saucer's pressurization drops suddenly and the two passengers climb out dizzily, with slight headaches. "A memorable day," Falco writes in his diary. "I still have the impression I'm dreaming."

There is another dive the next morning at dawn. Falco and Mollard take the saucer down to 60 feet, with a nylon line attaching it to the *Calypso*. Here is Falco's account:

"The saucer lands a bit heavily on the sand. To lighten it I start the water pump. I'm able to make the saucer rise to avoid coral outcroppings by shifting the mercury completely to the rear and accelerating the motors to high speed. I make several ascents and descents around obstacles on the bottom this way. I can easily follow fish, and I manage to make headway against an underwater current. I notice that the rudder has little effect, and I have to use the nozzles to turn. When we get back to the *Calypso* everyone is delighted."

The crew that witnesses these first tryouts are among the most qualified in the world to appreciate the importance of what is happening. Some of them have been involved in underwater exploration for five or six years. They have experienced the wonder of diving with scuba gear. But they also know its limits and dangers. Now they know that a new frontier is ahead: this is the beginning of dives to the deepest reaches of the sea, far beyond the 120 to 150 feet maximum possible with a scuba.

As for Falco, he has easily made the transition from diver to pilot of the saucer. He has felt no apprehension aboard it.

"I was in it so often during the time we were working on it in Marseilles and I had worked on it so much myself that I knew it down to the last detail," he says. "When I began actually diving with it, I had the impression that I already knew what to do"; this, despite the fact that the saucer is a complicated machine to run. On the left side, two hydraulic controls turn the two nozzles through arcs of 270 degrees. Next to that there is knob with which the water flow through the nozzles is controlled, making the saucer turn. On the left there is also the control for a telescoping arm that carries a 3-kilowatt light for illuminating the bottom, a fish, or a detail to be observed.

On the right-hand side, under the pilot's seat, the dashboard includes controls for starting the two-speed motor, controls for shifting the mercury to the rear or to the front to point the saucer up or down, the lever for maneuvering the outside claw, mechanisms for running the water pump to empty the ballast tanks, and lighting switches.

Also on the right is the lever for dropping off the two small ballast weights of 50 pounds each, one being dropped when the bottom is reached and the other when it is time to surface.

On the starboard side, aft, a control releases a 400-pound safety ballast.

Besides these controls for piloting the craft, there are two telephones, one for

Hoisting the saucer to the *Calypso*'s deck after a test for watertightness.

use on the surface, the other for diving, and a sonar, plus switches for the photographic equipment.

That same day there is another two-hour dive to 157 feet. Then the *Calypso* sails to Guadeloupe. From the 19th to the 25th of October there will be six dives near a small island called "pigeon's island." They last from three-quarters of an hour to two and three-quarters hours and go down to a depth of 210 feet.

During these dives Falco takes some excellent photos thanks to the camera and flash attachment incorporated in the saucer's front end. They constitute the first photos ever made at depth by a submarine. He takes along Professor Harold Edgerton of the Massachusetts Institute of Technology on one of these dives. Professor Edgerton was later to invent an electronic flash for the saucer.

Explosions

At 10 A.M. on October 25 the saucer dives once again with Falco and Mollard aboard. Commander Cousteau dives with a scuba to shoot some rolls of film of the miniature sub. Then Falco heads for greater depths and lands on the bottom at 210 feet. That is the deepest the craft had ever been.

Maneuvering is difficult, and every move calls for trying something new to avoid an accident. Falco is unable to get the saucer light enough, and it runs for the bottom in a barely controlled glide, landing heavily and crushing a bed of coral. Some of the coral branches hook around the nozzles and imprison the saucer. Falco finally manages to break loose after ten minutes by ramming backwards and smashing the coral, which leaves white, powderlike traces floating in the water.

He then heads up along a slope and meets Commander Cousteau, who wants to film the saucer as it casts off the safety ballast. He gives instructions to Falco by writing on a plate. Just as the big weight is being dropped there is the sound of an explosion, followed by another. Cousteau looks in at the porthole. He is worried at first, but inside the saucer all is well. Although uneasy, Falco and Mollard remain calm. They do not understand what has happened, and wonder whether other explosions are coming.

On the surface the crew hurries to get the craft out of the water as fast as possible. They check several times by telephone to ask if everything is all right. The saucer touches down on deck with a slight bump.

"Open up fast," Mollard tells Falco calmly, "we're on fire."

Their ears pop as the hatch is unscrewed and opened. The saucer is en-

One of the first tryouts of the saucer at Pigeon Island near Guadeloupe. Commander J.Y. Cousteau gives his instructions to Albert Falco at the controls by writing on a plate.

veloped in smoke. A battery is on fire, but fortunately it is on the outside of the craft. The crew gets the two men out and the saucer is dipped back into the water to extinguish the fire. The cause of the accident is soon found: the copper battery containers had short-circuited.

300 feet down

Cousteau asked Mollard to go back to France to have plastic battery containers made. Mollard rejoins the *Calypso* at the Cape Verde Islands at the end of November 1959 with the new containers, which are mounted on the saucer. They are tried out off Santa Clara Island.

At 10:55 A.M. on December 3 the craft is put into the water. At 90 feet a

A view of the Caribbean 60 feet down.

strong current carries it some ways out to sea and Falco lands it on the bottom at 180 feet on a 45-degree slope. He is unable to lighten the craft fast enough, and seeing that it is in danger of being swept along down the slope he casts off the ballast and surfaces.

But the saucer has to be tested at greater depths. That same day it goes down empty on a cable to 600, 900, and then 1200 feet.

December 5 is the day for the first very deep dive with occupants, and Commander Cousteau decides to accompany Falco.

The hatch is closed at 11 A.M., and at 11:13 the sub is 60 feet below the surface. Captain Alinat dons a scuba and goes along to shoot some film. The saucer is on a bottom of volcanic sand with a slope of 30 degrees.

The saucer being lowered by the *Calypso*'s crane.

At 11:27, Falco starts down the slope slowly. At 11:40 he stops and manages to hover 15 feet from the bottom while continuing to glide on down, reaching a depth of 300 feet. This is the first time that the craft has descended to that depth with passengers aboard.

The diving saucer traveled by plane when it was transported to the United States in this American Globemaster.

Suddenly there is the blast of an explosion and the saucer gives a jump. Then there is a second explosion. The depth gauge on the dashboard is stuck, but Falco notices that the plankton outside his porthole seems to be rising rather fast. In fact it is the saucer that is sinking. He casts off the second weight. The saucer seems to hesitate for a few seconds, then starts to sink again. The plankton is rising faster than ever. Cousteau releases the big safety weight, and this time the saucer surges toward the surface as the two occupants hold onto their seats. The saucer is going like a shot, and a few seconds later the depth gauge marks 0. A diver jumps from a waiting launch and attaches a cable to the craft while fire crackles in the water and in the air like fireworks. The batteries are burning again from a short-circuit.

"It was an awful thing to see and hear," Falco says. "It was impossible to open the sub and get out because water would have rushed in and drowned us before we could make a move. We heard Alinat say over the telephone, 'We can see smoke. We're on our way.' "

Cousteau replied, "There's nothing we can do for the moment, so we'll just wait it out."

The saucer was still floating, but on the outside the sound of crackling was louder than ever. Whole pieces of metal were melting.

Cousteau had brought along a snack that he had intended to eat with Falco while they made observations on the bottom. They had been too absorbed at that time to think of eating, so Cousteau decided that the time had come. He and Falco got out their food and uncorked a bottle of red wine as the batteries burned.

At last the saucer was lifted out of the water and set on the *Calypso*'s after deck. The two occupants got out amid a cloud of acrid smoke as the batteries continued to burn down.

The saucer was stored in the hold until the *Calypso* reached Marseilles, where the battery containers were changed for good. From then on, it never had any technical problems. Such minor incidents apart, it fulfilled all the hopes of Cousteau, Falco, and Mollard.

All the while, Falco was learning to pilot the craft:

"At 300 feet down in clear water you can see about 120 feet ahead," he says. "But beginning at around 400 feet you can see only as far as the floodlights reach, or about 45 feet. . . . The saucer acts a bit like a billiard ball. It has a tendency to turn on itself. It's not enough to point it in the right direction to get where you want to go. You have to predict its strange behavior and calculate instinctively how long it will take to get through a maneuver. To turn left, for example, you cut the flow to the left nozzle by turning a knob that squeezes a conduit. Turning the saucer is a little like driving a tank, except that, unlike a tank, the saucer keeps on going straight while leaning to the left. The actual rotation to the left only happens a few seconds later. You have to think ahead constantly. That's why this sort of piloting is tiring: it creates constant mental tension. You must never get excited and never react too strongly with a craft so susceptible to the least incident. All you have to do is brush against a coral fan and it slows down or starts to turn. Running along a submerged cliff face I sometimes brush against it and the saucer reacts immediately by turning, rising, or diving."

Deep dives

In 1960 the saucer is overhauled in Marseilles and put aboard the *Calypso* for a diving mission off Corsica. For the first time Falco, with Commander Cousteau along, reaches the depth of nearly 1,000 feet. Here is his account of that dive on February 2, 1960, in the Gulf of Ajaccio:

"After checking out the saucer thoroughly, we separate from the *Calypso* in 60 feet of water in a heavy swell. Ahead of us is an underwater ledge; I speed up to clear it and come down heavily on the bottom at 90 feet. There the Commander and I go over the craft again. Outside our portholes we can see divers going by making a film. They accompany us to 135 feet and then regretfully wave goodbye to us before running out of air. We set our course for the deep. We pass through a sandy valley between two rocks 60 to 90 feet high, then follow a slope where the bottom appears to be getting softer; at about 550 feet the slope levels off, and since there is very little daylight here we turn on the floodlamps and continue on down. We stop when the pressure gauge shows 600 feet, my heart beating a little faster than usual at the idea of being down this far for the first time. I check over the craft once again. The Commander asks me if all systems are functioning normally.

" 'Yes sir.'

" 'All right, let's go.'

"I accelerate and shift the mercury to the front, making the saucer head down an incline of about 30 degrees. A little school of pink fish glide by and disappear into the dark. Our passage upsets long white and pink worms planted in the bottom like so many cornstalks, and a dogfish comes up to look us over without raising the slightest cloud of sand. The Commander would like to film all this, but we are in an inconvenient position during the descent.

"We descend to 985 feet and there we stop. There is absolute silence. I can hear my heart beating. Suddenly we feel two bumps. The batteries again? The Commander reassures me:

" 'No, that must be hydrogen bubbles escaping.'

"He asks me to try to land on the slope. The saucer is perfectly balanced, and I maneuver it easily to the bottom. The safety weight touches first, and we stop the motor. The total silence makes me feel anxious, and I feel like making some noise to break it. We've already been down two hours, according to the clock, which reads 3:30. The Commander decides that it's time to surface. Going back up our position is better and after getting the telescoping light into position Cousteau films everything interesting that goes by. I do my best to avoid touching the bottom because every time I brush it there is a small avalanche that could eventually be dangerous for us. Once past 600 feet we notice again the octopus holes we had seen on the way down, as well as sea urchins the size of bowling balls. Going from one plateau to another past sponges and gorgonia, we stop at about 150 feet to have a snack. We don't stop long, however, for the carbon dioxide content in the cabin is at 2 percent, giving me a slight headache. I cast off the surfacing ballast, raising a sand cloud beneath us. It is 5:30 when the saucer's twin water spouts break the surface."

Serving research

The saucer is soon spending most of its time on scientific research, enabling biologists to observe marine life at depths never before attained. The first French scientist that Falco takes down is Professor J.M. Pérès, director of the marine research station of Endoume. Here is the account of that first experiment of its kind in the history of oceanography and biology. It took place on February 3, 1960, off Ajaccio:

"The saucer lands at 180 feet, perfectly balanced. Then we head for the bottom. At 275 feet we find little isolated rocks. At the professor's request the saucer is stopped and the floodlamps illuminate one of the rocks.

"Professor Pérès dictates into the recorder a description of all the species attached to the rock. Using the nozzles, we pivot around from left to right.

"We follow the slope down to 430 feet and stop several times so the professor can determine exactly which species we pass.

"Heading back toward the surface, we encounter a wall of rocks that we fol-

Two views of the Cassidaigne Canyon taken from the saucer at 600 feet.

low diagonally. Along the way we note a bouquet of yellow coral, some pink fish, and giant sea urchins with short needles.

"We ascend to a plateau at about 60 feet below the surface. Then, with the batteries down to 115 volts, the surfacing ballast is released and we break surface in a force two sea. We try unsuccessfully to pick up a launch or the *Calypso* on the radio. But after a few minutes a launch finds us anyway and tows us to the ship, where the crane hoists us to the afterdeck."

The canyon of Cassidaigne

During the month of December 1960, the saucer and its pilot were put at the disposal of several scientific researchers who wanted to study underwater canyons in the Mediterranean with this new tool, and in particular the east canyon of Cassidaigne.

Given the scientific program established and the difficulties in observing this area, these dives took all of Falco's virtuosity in piloting the saucer. Here, for example, is how a dive on December 4, 1960, went in the canyon of Cassidaigne with Mr. Jacques Laborel as passenger:

"Today we are supposed to bring back as many specimens as possible with the saucer's claw, particularly one of the large anthozoan corals that grow on rocks about 420 feet down. We are launched above a plateau with a depth of 250 feet. We head west to find the rocks. After 15 minutes we see the rocks on a slope of about 20 degrees. They are covered with sponges, coral, gorgonia, and a few bouquets of yellow coral. There are also some anthozoan corals three- to six-feet tall, virtual underwater tamarisks. I start by gathering some sponges. Maneuvering is very difficult because the slope is steep and the bottom rocky and hard. If I brush against the slightest coral growth while attempting to seize a sponge with the claw, the saucer swerves away, making me start all over. That means backing off and returning to the same spot, finding the sponge again, and edging up to it by playing the nozzles just right to bring the claw precisely over the sponge. It's like playing with one of those carnival gadgets where you try to pick up a fountain pen or cigarette lighter with a toy crane. Usually you end up with nothing. Fortunately I'm luckier with the saucer, especially today: after two hours of work I have a good collection of samples, including a basket full of sponges, sea urchins, and gorgonia. The hardest was to get an anthozoan bush because once I had it in the claw I couldn't pull it out of the bottom. I tried coming at it full speed, using the saucer like a bulldozer, but I missed and flew over it. The second time around the bush was lying loose and I got it in the claw triumphantly. I brought the claw arm back against the hull and when we broke surface a few minutes later the bush was still there, framed by two spouts of water from the nozzles, like a water show at Versailles."

Rech Lacaze Duthiers

The saucer made four dives in the canyon of Port Vendres, which is designated "Rech Lacaze Duthiers" on nautical maps. In Falco's reports can be seen not only the wonder men feel when they contemplate the marvelous spectacle of the depths, but also accounts of some dramatic incidents.

The first dive, on January 10, 1961, with Mr. Daniel Reyss of the Arago Laboratory of Banyuls sur Mer as passenger, was extremely rich in terms of biological research. The greatest depth reached was 722 feet:

"After encountering ascidiums, annelides, and empty shells on the plateau, the saucer dives into the canyon to cross it on an easterly heading. At about 590 feet small gorgonia can be seen on the black bottom on the edge of an immense ledge. On the vertical part of the ledge there are small red corals, sea urchins, and, farther on, a virtual bouquet of yellow corals and magnificent gray and white sponges in the form of a funnel.

"As we are admiring the spectacle and heading up the slope that followed the ledge, there appears a large mass swimming toward us. It is only after it is within 30 feet of us that I recognize it as an enormous sea grouper weighing at least 110 pounds. I start filming it and it keeps on coming, doubtless blinded by the floodlamps, and bumps into us head-on. The collision jostles us inside. It makes a swipe at the saucer with its jaws, brushes against the hull, and all we can see as it passes beneath us is its raised dorsal fin. It doesn't look pleased about its sudden encounter with the steel of the saucer and slowly disappears toward the depths. My companion and I are sorry the visit was so brief. Absurdly, we had thought that it might be able to get used to our presence. The rest of the dive, to 720 feet, was not particularly fruitful, revealing only a few rocks and lot ofshells and dead white corals. The liveliest things we meet are a little shark with yellow eyes and some sea urchins."

The S.P. 350 during a night dive at Guadeloupe.

At minus 980 feet

On January 16 there is a new dive with Henri Laubier to 980 feet:

"At 6 A.M. the *Calypso* slips out of Port Vendres on a calm, windless sea. At 8:15 the saucer enters the water and touches the bottom at 600 feet. There is a strong current and making headway is difficult. At times we hang motionless, even on a 35-degree slope, and it takes over ten minutes to go 15 feet to visit a first rock covered with white and yellow corals and white sponges in the shape of ears. There are also some tiny branches of red coral. Rocks are all along our route to 915 feet, where fortunately the current diminishes. We are lucky to come across an enormous rock covered with bushes of white and yellow coral and antozoan trees, especially at the top, which has a 35-degree slope on one side and is flat on the other. On the other side we find some lobsters and a big-eyed sea-bream with a black spot in the middle of its body. It and several redfish stay about 30 feet from us without approaching. Unfortunately I am unable to shoot footage, my large floodlamp having gone out. We reach a depth of 980 feet, but we have stayed down ten minutes longer than we had agreed with the *Calypso*.

"I cast off the surfacing weight and as much ballast as possible to go faster. The radio is working well."

A diver attaches the saucer to the *Calypso*'s crane to hoist it aboard.

There was another dive into the canyon on January 17 with Daniel Reyss that went to the saucer's maximum depth: 991 feet.

"The saucer lands on a bottom of mud and shells at 525 feet; the slope is gradual, the current of about one knot carries us south. I notice traces of trawling in the sand. There are a lot of fish (red mullet, rays, red gurnet). At 820 feet I hop over a range of rocks and maneuver between them and the muddy slope. Deeper, at about 885 feet, we meet a strange gray-black fish with a white belly having the body of an eel, the mouth of a pike, and a small rounded tail. I do my best to approach it, but the current is against us and the fish keeps its distance, circling us. Taking advantage of our floodlamp to find prey, it gobbles up two or three fish. It's a virtual monster weighing about 20 pounds. The water is cloudy, making navigation difficult. We keep heading down and, not seeing the fish again, stick to our course by zigzagging between boulders. The pressure gauge indicates 950 feet, and at that point the water clears. About 30 feet ahead, standing out in the night, we can see the long precipice that marks the end of our dive at 985 feet. I advance in the blue and turn around at the precipice, where there are numerous white corals. I approach closer to film them when suddenly several large red shrimp shoot out of a bouquet of coral and dance in front of our portholes before disappearing along the cliff face. Farther down, at about 1015 feet, the white corals are denser, but unfortunately that is too deep for us and we regretfully leave this nether world."

Swept away by the current

The saucer's 47th dive is on January 18; the passenger is Mr. Henri Laubier and it will also be a deep dive:

"The *Calypso* launches the saucer on the plateau to the north of the Rech Lacaze Duthiers canyon. When we reach the bottom I hit the same strong current toward the south as on previous dives. It carries us along a 20-degree slope. The water is cloudy, giving us 10 or 12 feet of visibility at most. I feel uneasy, for with this current we are making at least 2 knots and I'm afraid of running into an obstacle. We go along for nearly an hour without seeing anything interesting except some sea urchins that leave deep traces in the mud. The bottom levels off and the water clears a bit. The sand is scattered with the debris of dead corals and urchins. A strong east-southeast current suddenly begins pushing us faster; I can hardly keep on course and I'm worried as we start to climb the east side of the Rech. I never like climbing up slopes because I'm always afraid that I'll run up under an overhang without noticing it. I'm looking intensely ahead and am covered with sweat. It's just as well that my passenger is unaware of the danger.

"I miss some rocks by quickly shifting the mercury and have to reverse just in front of a vault. I would like to cast off some weights, but I don't dare. I'm afraid that we will get stuck beneath the rock. We can see nothing all around, and my passenger becomes uneasy at no longer seeing the bottom. I decide to head for the cliff face, taking a heading from the compass. I don't want to cast off ballast without knowing what is above us. The pressure gauge indicates 590 feet. After ten more minutes I see ahead a muddy dome scattered with large pebbles. I brush over the top of it and note that the current is now toward the south. The water becomes clearer. I give the saucer its head, and we descend a steep slope of about 35 degrees. At 787 feet the current stops short. I see a rocky ledge in front of us with a void on the other side. I slip us down foot by foot, making sure not to start any avalanches by touching outcroppings.

"Some magnificent white coral bushes are attached to the cliff face. We shoot a lot of still and movie film of them. Large plankton dance in front of our portholes: undulating, transparent worms with two small red horns, phosphorescent squid, and red shrimp form a sort of halo around the saucer. Above us a white mass passes by and I point the nose of the saucer up. It is a funny-looking triangular shark of a sort I have never seen before. At about 920 feet I can no longer see the bottom, with visibility limited to about 30 feet owing to all the particles in suspension. There are fewer corals between 920 and 980 feet. At last we touch the foot of this cliff that seemed bottomless.

"I descend another 20 feet and note immense rocky blocks in the mud. I shoot some film of them, but at that moment the saucer gives an unusual vibration and I get nervous. I cast off the surfacing weight and we watch those rocks disappear once again into the night."

The two eels

The 57th dive was also into the Port Vendres canyon. It gives us some idea of the magnificent spectacle outside the saucer's portholes. This time the passenger is Pierre Drach, a professor at the Sorbonne and director of the Banyuls Laboratory.

"The current threw us off course during the descent, and we land about five feet from the edge of the big cliff. I turn the saucer carefully and we look down a precipice. The professor is delighted and asks me to go closer to an overhang full of white coral. I stop us about 18 inches from a marvelous setting suspended from the face. While the professor takes notes, I position us slowly, using the claw arm to push us gently backward without raising any mud. We are perfectly balanced and remain easily in the same spot. Plankton dance all around us, along with silvery little fish swimming vertically. Blinded by our floodlamps, they run into the coral, and several are caught on its points. There is a sort of silver powder falling all around us. Farther down, at about 820 feet, a rocky promontory extends toward the south. I go over to it and we see a sea spider that we film. I land the saucer at 950 feet near a rocky mound and gather corals with the claw. But unfortunately the batteries are weak and it is hard to maneuver. All at once, two enormous eels weighing about 65 pounds each pass in front of us; I did not think they existed at such depth. They pass by several times while I am gathering samples. I try to catch one with the claw, but it shakes loose and makes the saucer spin around once. We start back up."

However marvelous the miniature sub may be, its use sometimes results in surprises, as Falco's story shows:

"The saucer descended along the cliff at Nice to 885 feet. I was teaching André Laban to pilot it. It was the first time he had been down so deep with me, and we were starting to run along a vertical cliff face that went down to 900 feet. All at once a stream of water hit me in the back and I gave a jump. The leak came from the hatch. At the same moment there was a large flame inside the saucer. The water had hit the switches for the electronic flash. I quickly cast off the surfacing weight and we shot up.

"When I examined the craft afterwards I discovered that it was not the hatch that leaked but a joint in the pump. The spurt had been so strong that it hit the hatch first and then bounced toward me. It had been scary."

A new world

Albert Falco has made over 600 dives with the saucer to date. Some have lasted several hours, many have been down to between 900 and 1500 feet. He

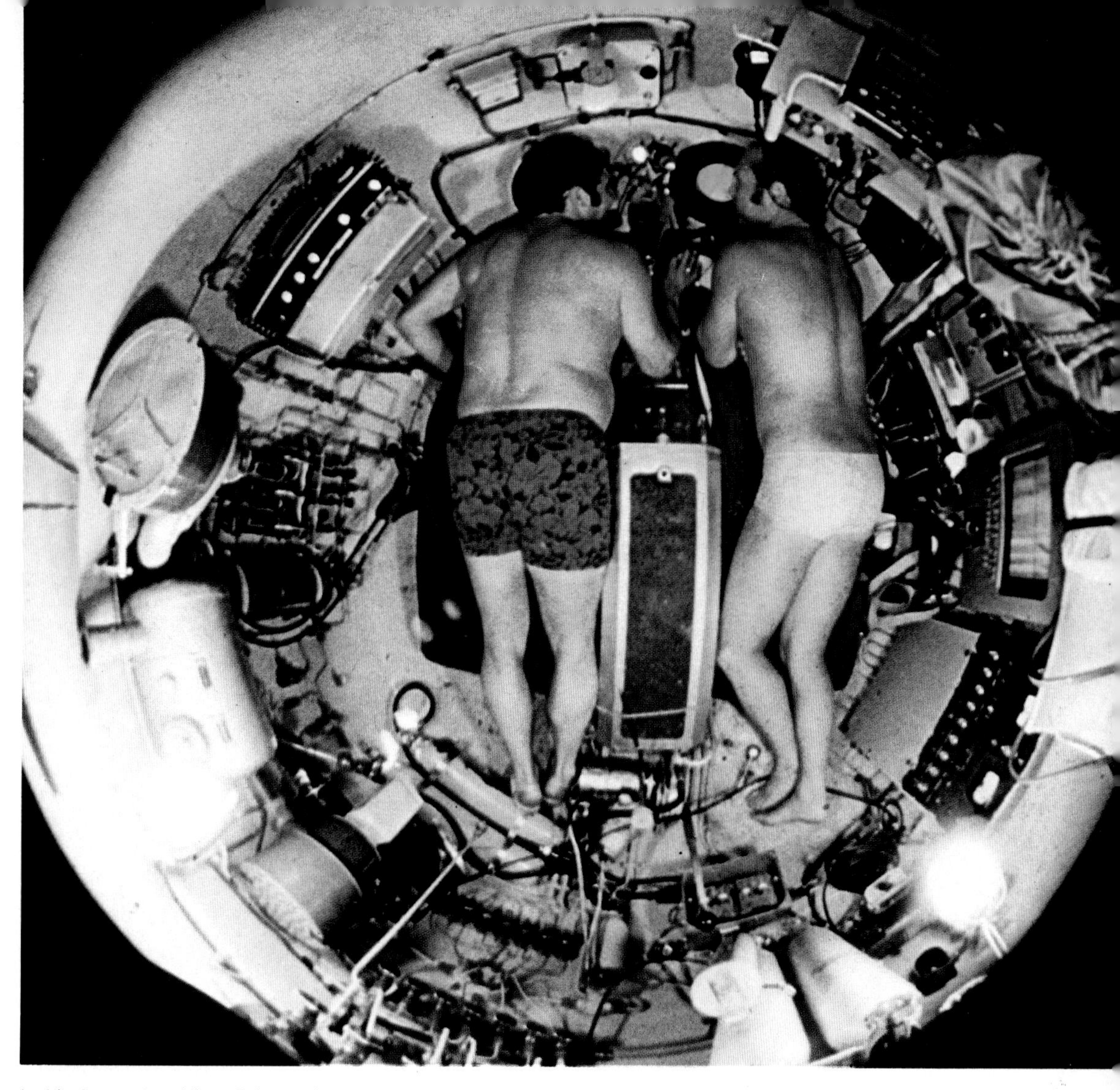

Inside the saucer: Albert Falco and a passenger.

probably has seen more of the underwater world and at greater depth than any other man. Thanks to his use of scuba gear and the saucer, his knowledge of the sea is unique. He admits, though, that occasionally his visits to the bottom can be disappointing.

"Most of the time there is nothing to see down there. You can go for miles over nothing but mud or bare rock. On a run of two miles, for example, you can go for a mile and a half without seeing a living being, then suddenly come upon a spot 300 yards long full of coral, shrimp, and squid.

"I have followed the pipeline that starts from Cassis and leads to the Cassidaigne canyon 1150 feet deep. Starting at 210 feet down there is nothing but mud, but there is still evidence of life. The mud is inhabited by gobies, crabs, shrimp, and the like, but they are all very small.

"Sometimes you find an area full of antedons or brittle stars, Echinodermata with multiple arms that they wave wildly. It makes a beautiful sight to glide above a carpet of antedons stretching for 500 yards. These strange animals eat by sticking their arms out into the current and catching tiny organisms. You can see thousands of them on the bottom, and a few rise and swim in the shaft of light from the saucer. It's an extraordinary spectacle.

"Then the bottom becomes barren again until you come upon an area inhabited by sea urchins that live at great depth like *Stillocilaris* or *Dorocilaris,* with a small body and long needles. There are also the very large urchins called melons: *Echinus melo.* Their presence indicates large rocks nearby. You can also tell that rocks are near by the number of empty shells accumulated in an area.

"Sometimes the saucer runs right into a school of fish, for instance sardines. . . . Once when I was going along a slight slope between 300 and 360 feet down I came across thousands, even millions, of hermit crabs running through the mud in herds. All along this slope there were also trumpeter fish, little pink fish that resemble triggerfish. There are two sorts of trumpeter fish, a small pink one with a normal mouth and another with a narrow, pointed beak. You can see them often in the Monaco Museum's aquariums.

"Farther down, rockfish were on the bottom. They often live at great depth.*

"Very rare madrepore corals were attached to a large rock, both white and yellow ones *(Dendrophylia);* they were one of the primary interests of Professor Pérès.

"Gorgonia and small sponges were attached to nearby cliffs like packages of cotton. Wherever there is a large rock at great depth, there are always sponges attached to it. Depending on the region, they will be small white or yellow sponges, or occasionally large ones with horns that look like ears."

* Found only at depths below 600 feet.

Cassidaigne has marine scenery as typical and varied as any of the provinces of France. Previously it was known only through soundings and dredgings, but no man had ever looked upon it and compared its multiple aspects. It is truly a new world that man is getting to know.

The S.P. 350 diving saucer has just been launched. A diver detaches the last cable linking it to the *Calypso*.

6

in the shadows of the sea

CAUGHT IN A TRAP—UNDERWATER COLLISION—A FORMER LEVEL—SECRET RICHES—SHARKS AGAIN

Exploration in the diving saucer revealed not only fish and coral, but also the geography of the depths. Previously, man had only abstract information about the bottom of the sea: it was summed up in a few lines drawn on maps. The lines themselves simply linked up points whose depth was known thanks to soundings made in several different areas. Thus such maps could give only an approximate idea of reality. The saucer, however, enabled men actually to see the bottom. The story of Professor Sheppard illustrates how much use of the saucer has revolutionized the study of the bottom of the sea.

Professor Francis P. Sheppard is an American geologist who worked at the famous Scripps Oceanography Institute in La Jolla, California. For 30 years he studied a canyon in the Pacific, outlining its shape, measuring its depth, and determining its geological composition through soundings and dredging. For 30 years he had a mental image of it, but he had never seen it.

"Professor," Cousteau told him, "we are going to show you your canyon." That was in January 1964. Sheppard got into the saucer with Falco.

"The professor was very excited," Falco remembers. "He was finally going to see the canyon he had devoted 30 years of his life to. During the dive he sang *Santa Lucia* with me in Italian.

"The pressure gauge indicates 590 feet when we touch bottom. Fifteen minutes later we are going along a slope of 15 or 16 degrees and see some large shrimp. The mouth of the canyon starts with a bottom of shifting sand hollowed out to form a hole 50 to 60 feet deep and inhabited by numerous torpedo fish.

"The saucer continues along the slope, with cliffs closing in on the right and left; they are composed of rock piles at first, then of massive stone. As we go deeper the sand flows along like a river.

"Professor Sheppard is enthusiastic as we pass through parts of the canyon that close above us like a vault. If we had a breakdown, it would be impossible to get back to the surface.

"At 600 feet the rocks form a slope of 20 to 30 degrees. There are a lot of scarlet fish resembling groupers, madrepores, anemones, and Japanese king crabs. The crabs use their two rear appendages to pick fan corals that they carry on their backs as camouflage.

"We reach the bottom of the canyon at 755 feet. In the clear water we can see sand, dead algae, gray, freshly broken rocks, and some soles about 40 inches across. We run along the cliff on a heading of 330 degrees, picking up samples of rocks and sediment. The slope becomes steeper.

"We arrive in the Scripps canyon and change to a 60-degree heading. We ascend 300 feet, crossing the canyon to see the other side, then go west to end at 900 feet on a heading of 300 degrees."

Professor Sheppard thought he knew his canyon well, but he could not have imagined it as it really was, with its vaults, corridors, and tributaries. It was a veritable underwater river, with all the details of landscaping you would see along a mountain river on the earth's surface.

The reality thus revealed in the floodlights of the saucer at first filled the American oceanographer with exuberant enthusiasm, inspiring him to sing *Santa Lucia,* but he soon became quiet as this was replaced by a feeling of admiration for what he was seeing—the unprecedented sights of a new kind of oceanography.

Caught in a trap

Falco took another researcher from La Jolla, J.W. Curray, to the same canyon, and the incident he had worried about the first time happened: the saucer got caught in a trap.

"I saw an anchor chain partly embedded in the floor of the canyon. I thought I would find a shipwreck a little higher. I followed the chain up, but a strong current pushed the saucer into a corner formed by the canyon walls. I broke out in a cold sweat: I couldn't go up because the walls leaned together above us. I tried to turn the nozzles gently, but I was afraid that they might speed us up too fast and bang into the rock. If they got damaged, we were done for. I didn't want to cast off the ballast because if the saucer climbed it would get stuck between the walls above and it would be impossible to get it out.

"Finally I was able to reverse the nozzles without ramming them into the cliff, and tried to accelerate. Nothing happened. We were completely jammed. For a moment I had the feeling that it was all over, that we would stay blocked in the canyon. Then I had an idea. I got up, went to the center of the saucer, and started rocking. I could feel it moving. A few seconds later we were in the middle of the canyon and moving freely. We had had a close call."

Underwater collision

In March 1963 the *Calypso* sailed again for the Red Sea and the Indian Ocean with Falco and the saucer aboard. The area was familiar to the crew: the Jubayl Straits, the Shab Ali reefs, and especially the little sharks that cruise through those waters. But the sea was heavy, and a sandstorm made it impossible to use the saucer. Finally on March 10 Falco tried a first dive.

He descended among gorgonia and Alcynaria coral, saw some giant clams, and landed on a plateau at 115 feet. Then he moved again and landed heavily on a sandy, 35-degree slope. The current carried the saucer down to 490 feet, where it stopped against some dead coral at the edge of a cliff.

Falco had trouble getting started again in the strong current, but managed to do so by shifting the mercury all to the front. The saucer started toward the bottom at high speed. Along the way Falco noted virgularia, yellow madrepores, and yel-

Double page following: A team of divers explores the coral of the Red Sea.

low and white gorgonia. The current became stronger and more irregular, alternately rising and descending. All Falco could have done in such conditions was to follow the cliff face. After a while he managed to leave it and stop beside some coral that the photographer, Goupil, shoots some footage of. At 2:20 P.M. the batteries were dead and Falco headed for the surface.

"I'm exhausted," he says. "I've never had such a hard time with the saucer."

Falco made numerous dives with the saucer during that cruise, several of them memorable.

To the south of Gardaufui, five or six miles from the island of Socotra near the Indian Ocean, Commander Cousteau sounded a cliff that went down to 900 feet, cut across by a plateau at 600 feet.

"We went down with the saucer with the idea of landing on that plateau," Falco says. "We were about 60 feet above it when I noticed little black spots all over the sand. I thought it was spatanginas, those little urchins that push their way through the mud like bulldozers. But when I cast off the first weight at about 25 feet from the bottom, I realized that it was thousands of mating crabs. They were grouped two by two and piled all over each other. A carpet of crabs covered the bottom for a length of 200 yards.

"Heading south, I came to a crest. Beyond it the cliff dropped vertically. I went down slowly into the void. I could see large splits in the rock like slashes of a sword. In the dark I could make out millions of crustaceans with phosphorescent eyes and big scarlet fish with large heads and black round eyes.

"Lower, on a ledge about 25 feet wide, I saw a strange flat fish, round with a small tail. It was greenish yellow with a bit of brown and orange. We went up to it, and to our surprise it rose, swelled up, and began to waddle away, stretching out its feet to walk on the rocks. It waved a little antenna above it with each step and it measured about 15 inches long. We shot footage of this nightmare from all angles, and I managed to catch it with the claw after four tries.

"Toward the end of our dive I saw a shark on this same cliff, an enormous one 20 feet long. It lived at great depth and had no dorsal fin. Blinded by our floodlamps, it came at us and then turned away at the last moment, running into the cliff and starting a little avalanche.

"It must have been attracted by the odor of the bits of tiger shark that I had put in the basket attached to the outside of the saucer before the dive. It came back three or four times. I tried to follow it as it ran into the mud and then into the rock. I was fascinated by it as we matched wits. It disappeared into the dark, then returned and jostled the saucer with a wave of its tail. Its mouth passed 20 inches from my porthole, and I could see its double rows of pointed teeth like the thorns on a rose bush. Its gills were enormous. The turbulence from it had upset the saucer and I had to get it back under control at around 830 feet. I was able to shoot movie and still photos of it before it finally disappeared below us, too far down to follow. The pressure gauge indicated 935 feet. The saucer's batteries were dead and it was no longer responding to the controls.''

A former level

Falco and Commander Cousteau made over ten dives to note the appearance of the bottom at different depths. First they examined the shallow reef area with its corals and multicolored fish, the umbrellas of the big Acropora corals, then the dead coral trees at 100 to 120 feet.

With each dive they noted at about 300 feet in different areas a ledge or balcony that could only be a former level of the Red Sea and Indian Ocean. The water was nearly always much colder at this exact point.

Descending from this ledge, they entered large vertical valleys with a strange decor seemingly carved in the dead coral. The shapes sculpted in the cliff face resembled columns or immense jars. At certain levels the piles of dead coral looked like old bones. But at great depth the Red Sea was not very rich in marine life. Beyond the ledge, at 300 feet down, there was little likelihood of encountering a fish, except the odd tuna or shark passing through.

Falco and Commander Cousteau dived again, with Cousteau taking the controls on a bed of sand at about 215 feet. He maneuvered around blocks of coral, brushed against a gorgonia at the edge of a cliff, and shifted the mercury downward to begin the most difficult part, the descent along the face. He changed the water ballast several times, spraying Falco. After a few minutes the saucer assumed the right course.

It was a new experience for Falco to ride along and admire the scenery. Unfortunately the cliff was bare, and there were no fish.

At 965 feet the muddy bottom was scattered with a few large boulders. But the sea was empty, and the Commander lightened the saucer and ascended. Here we pick up Falco's account:

''At about 590 feet we can see a sort of immense cobweb hanging in the water. It is composed of floating particles linked together by filaments. The Commander has to make his first difficult maneuver. After a second's hesitation, the

Albert Falco and J.Y. Cousteau get into their suits for a night dive.

On the *Calypso*'s afterdeck preparing for a night dive: from left to right, Armand Davso, Albert Falco, Jo Thompson, Yvan Giocoletto and J.Y. Cousteau.

The saucer 360 feet down on a night dive. In the foreground, a rock covered with red coral and sponges.

saucer heads straight for the silvery net and goes through it. It's a wondrous sight, and I shoot with both cameras at once. The cobweb comes apart, and when we turn around there is not a single thread of silver in the sea. I can't imagine what it was."

During the spring of 1963 Falco makes a great number of dives in the Red Sea with the saucer, which had been equipped with an ultrasonic underwater telephone. It can communicate with the *Calypso* 1,500 to 2,000 yards away from a depth of 200 feet. The new scubas of the divers were also equipped with them.

Secret riches

Falco was fascinated by the hours he spent in this supernatural decor, contemplating coral while surrounded by sharks. What with that and the resulting fatigue, he sometimes lost track of time.

"Sometimes the Commander thought I wasn't coming back up," he says. "It was so exciting that I spent hours under the surface."

The exoticism of this tropical underwater world was increased farther down by a fantastic architecture of coral in the shape of organ pipes or giant columns, which we have already mentioned. The whole was plunged in darkness and made its appearance suddenly in the saucer's lights. Every night Falco brought back from his dives mental images that he was perhaps the only man in the world to possess. He was thus the depository of great and secret riches.

Here is Falco's account of one of his exploratory dives in the Red Sea:

April 5, 1963. "*The Calypso* is anchored off the Audi Seli reef and I notice the silhouette of a shipwreck with its starboard side sticking out of the water. The launch is lowered with two divers, Davso and Bonnici. Back on the *Calypso* 20 minutes later, they say there are a lot of fish and particularly a lot of sharks. The drop-off is steep, as usual.

"The saucer is launched at 2:25 P.M. Pierre Goupil, the cinematographer, goes with me, and right away we see ten sharks, about 20 bludgers, pompanos, and tuna. We are surrounded by hundreds of fish; I've never seen so many. They stay with us down to 300 feet, but leave us when we reach the ledge marking the sea's former level.

"Still, one of the pompanos hangs on. It stays beneath the hull and comes up to the porthole from time to time, finally going back up when we reach 600 feet.

"The cliff face is composed of a series of small, prominent steps without much growth on them. The columns are less well defined. At about 785 feet there are mounds of sand and debris of dead coral. Life is beginning to disappear. But suddenly as we head north I see two large luminous points. Although the batteries are weakening, I manage to get over to a block of coral riddled with holes. Down to the left there is an immense violet gorgonia fan, the largest I have ever seen. The coral is covered with long-legged crabs and little pink crustaceans with big claws. But we have run out of electricity aboard and the flash won't work. There's no more pressure in the hydraulic circuit, either.

"I can't make myself leave this beautiful sight without trying for at least some photos of it, though, and I slowly work the saucer over until we are about five feet away from the coral so Goupil can shoot some footage. A few inches in front of us, the crustaceans scatter in panic. Although I can't move a single hydraulic control, we pause a little longer to watch some magnificent scarlet fish with big phosphorescent eyes as dozens of gray-green fish hover around us. I get us over near the scarlet fish to film them, but as I shift my weight to swing the saucer around I'm sprayed with oil and the oil pump switch emits a cloud of dense smoke.

"Now we have to surface for sure, but first I have to get out from under a slight overhang on the cliff face. I work the pump switch manually and get enough power to maneuver to the right position for casting off the surfacing ballast. We are coughing as we climb out of the saucer on the *Calypso*'s afterdeck at 7:30 P.M.

"Then we sail for the island of Jebel Taïr.

"I am exhausted and Goupil has a headache. When I lie down to relax, I see colors from the sea before my closed eyes—red, yellow, blue, and orange—and then slip into a bottomless hole."

Sharks again

April 21. When the *Calypso* is moored at Shab Arab reef, Commander Cousteau orders a night dive with the saucer and several divers. The shark cage is lowered to the bottom at 60 feet.

At 8:30 P.M. Falco gets into the saucer with Corre and follows the cable down to the shark cage, where over 20 sharks are already gathered. A team of four divers made up of Pierre Goupil, Raymond Coll, Christian Bonnici and Gilbert Duhalde jump in with lights, but as soon as they enter the water the sharks show up. The divers have to use the shark stick to keep off the more aggressive ones. Following their air bubbles, Falco maneuvers the saucer to try to protect his colleagues, who have positioned themselves on the roof of the cage. The sharks are getting more nervous; there are now about 50 with more coming.

The cage is too small to shelter everyone. While Duhalde keeps the equipment in the cage, the others stay outside and beat off the sharks with lamps, flippers, and cameras.

In the saucer Falco is increasingly uneasy. He can see that the situation is getting worse by the minute. It won't be long before the sharks attain that point of excitement when they enter a sort of frenzy that nothing can stop.

Beside him, Corre panics and shouts to the divers as if they could hear him, "Get out of there before they make mincemeat out of you."

Falco moves the saucer as close as possible to the little group of men who are slashing right and left like knights in battle.

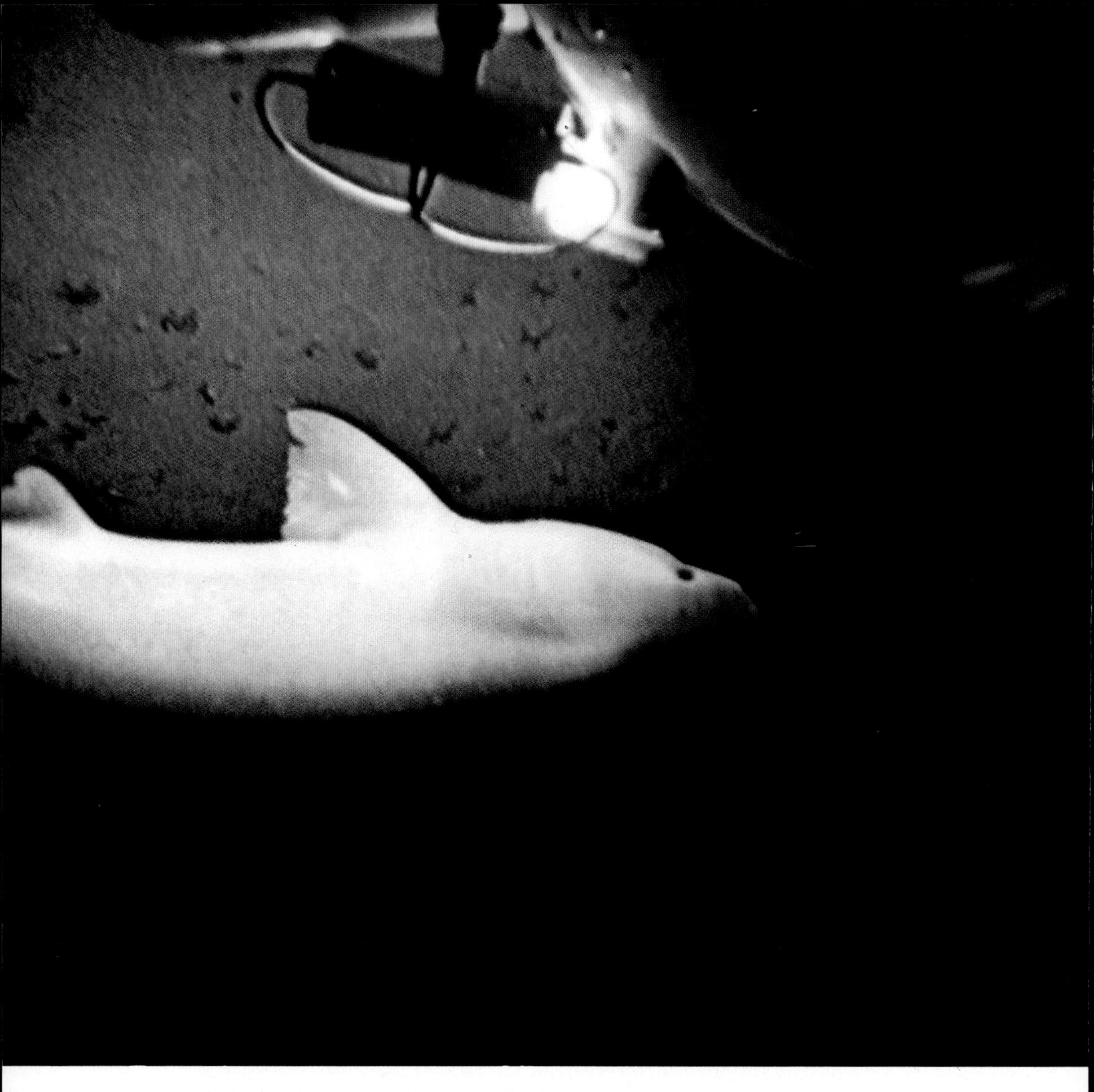

A shark, probably *Hexanchus griseus,* illuminated by the saucer's lights at 1,000 feet.

Surgeonfish, born armed with a scalpel at the tail, are sometimes a surprising pastel blue.

Raymond Coll finally leaves the group long enough to yank the cable twice, giving the signal to pull the cage up.

The cage rises slowly with the men hanging on to the bars as the sharks follow them all the way to the surface. The last ones are finally pushed away with the sticks. Man has won over the animal, but just barely.

Falco sighs with relief. "I've never sweated so much in that saucer," he says. But the night is not over yet and there are more adventures to come. Another dive is set for 2 A.M.

This time only two divers go down in the cage: Goupil for movies and Daniel Tomasi to make still photos of the sharks. The saucer meets them at about 90 feet, where the two divers are attacked not by sharks but by millions of little animals attracted by the lights. Goupil and Tomasi slap themselves on the legs and neck as if they were being stung by wasps or mosquitoes. At first Falco assumes that the animals are jellyfish. But in fact they are sea fleas. The divers are soon unable to stand the stings and are brought up. Their skin is bleeding wherever it was not covered by their diving suits. Their ankles particularly are covered with dozens of little holes.

This incident was to have a sequel.

The next morning Goupil tells Falco that his whole right side seems numb. Falco verifies the length of the night's dives and finds that Goupil should have made a decompression pause of at least 10 minutes, but in his haste to escape the sea fleas he made none at all.

There is not a minute to lose: the one-man caisson is prepared in the rear hold and the doctor orders six and a half hours of compression at 42 pounds per square inch for Goupil. When the pressure in the caisson reaches 28 pounds Goupil already feels better. But at midday in the Red Sea the metal caisson is hot and stifling. The crew takes turns sprinkling it with water.

At 2 P.M. Goupil begins vomiting. With the help of Jacques Roux, Falco hooks up a secon air line to increase the pressure. He talks with Goupil over the interphone to calm him. Goupil seems reassured and closes his eyes. He seems to understand that although the caisson is like a tight prison it is better to spend six hours in it than to be paralyzed for the rest of his life.

Falco opens the caisson at 5:30, after waiting an extra 15 minutes to be sure.

Goupil gets out feeling dazed and weak. He is wrapped in a coat and a blanket. He is hungry and thirsty. That's a good sign.

After a night's sleep Goupil is back in shape. But it was a good lesson for him that he is not about to forget.

CALYPSO

7

Shab Rumi, underwater village

DIOGENES—DREAMING IN THE SEA—FORGETTING THE EARTH—FOUNDING A VILLAGE—EXPLORATION—SHAB RUMI—THE HARDEST WORK—TUGGING—IMMERSION THE HOUSES—A FRIENDLY TRIGGERFISH—CINEMA—THE WRECK OF A HOUSE

Exploring the sea with scuba gear presented certain difficulties: stays on the bottom were limited, not only by the amount of air in the bottles, but also by the necessity for a progressive decompression. The pauses for decompression become excessively long in the case of dives of an hour or more.

As early as 1961 Commander Cousteau became convinced that the solution would be to set up an underwater shelter where the diver could live at the same pressure for as long as necessary; he would then have to come up only once. His blood would remain saturated with nitrogen for the entire length of his stay down.

Conshelf I: Immersing the underwater house from the *Calypso*.

Diogenes

To test this idea, Cousteau decided to keep a diver below the surface for several days. He naturally chose Albert Falco for the experiment. With another crew member, Claude Wesly, Albert Falco spent seven days at a depth of 30 feet in the Marseilles roadstead off the Frioul islands. His "house" was a cylinder of sheet metal dubbed "Diogenes" because its shape resembled the famous Greek philosopher's tomb.

Instead of being attached to a ship on the surface like a diving bell, the quarters of the first oceanauts were chained to 30 tons of lead that lay on the bottom. They entered through a door on the base of the cylinder. Inside, the air pressure was slightly greater than the water pressure at that depth.

Object of the experiment? To study the possibility of leading an active life for several days or weeks in an environment considered until then too hostile for extended and effective work.

Falco and Wesly left their refuge three times a day to work at depths of 65 to 80 feet. They were in the water for two hours in the morning, two hours in the afternoon, and one hour at night. They had to make no decompression pauses to return to their underwater domicile, as they would have if they had returned to the surface. In their cylinder they lived permanently at a pressure of 28 pounds per square inch.

Dreaming in the Sea

The experiment was hazardous. How would men stand this life in the sea, this isolation in an air bubble, prisoners of this liquid mass all round them? For the answer, we have to read Falco's diary. It shows at what cost the experiment was successful:

"I haven't dreamed for years; I catch up on lost time with a nightmare that I won't soon forget. Full of oppression, smothering, anguish, panic. A hand is strangling me, I have to get out, I have to surface. I wake up, I go to the observation hole, I look at the water, I check the dials one by one. Everything is normal, Claude is sleeping peacefully. I lie down, but can't get back to sleep. I feel completely isolated, trapped. We are condemned to stay under the water until the end of the experiment. We are no longer free to surface. To surface we would have to eliminate our nitrogen, and we can only do that with the help of those up there

Albert Falco and Claude Wesly sit at a garden table during tests. Behind them is the house used in Conshelf I.

who are observing us. I am filled with an irrational fear. To calm down, I think of my comrades up there and the Commander. Everything necessary has been done; this very moment someone is watching me on television. But I don't manage to reassure myself. I am obsessed by an absurd idea: what if the porthole broke and the water began rising? How fast would it come up? No doubt there would always remain enough air at the top of the caisson, we would have time to put on our scubas and get out. Yes, but even then we would have to overcome our usual reflex and not try to surface. I can't get back to sleep. The noise of the air at the water level is infernal, much more noticeable than during the day. It bubbles all the time like a giant kettle. It also resembles the noise pebbles make when thrown on the beach during a storm. Claude is really sleeping well. . . ."

During the Conshelf I experiment Albert Falco and Claude Wesly leave the house for a work session.

Forgetting the earth

Further in Falco's diary:

"I feel diminished. I have to move slowly, or I'll never make it. I'm afraid I won't be able to hold out. Everything is difficult to do, as if I had an obstacle constantly in front of me, a wall to get over. Today the work sessions in the water are really laborious. I feel better in the house, relaxing on the bed."

Two days later there are these reflections divided between hope and fear:

"The Commander has ordered us to rest. I think now that life beneath the water is possible for a long period and at greater depths. But what if we completely forgot the earth? It's true that I don't give a damn about what is going on up there anymore. Claude too. I feel completely detached. . . .

"Phone call from the Commander. He knows that we had some trouble yesterday. They sent us down an air bottle half-empty. Claude was playing around pushing tiny shrimp—there are millions of them here—onto Cerianthidae anemones in the 40-foot zone. The lashes of the anemones closed over the shrimp like tentacles. All at once Claude turned toward me and gestured wildly. It looked urgent. I took my mouthpiece out and handed it to him. He took a breath and gave it back. The house was fairly far away. We returned by sharing my air, without any panic. The last 20 yards Claude let go of me and made a run for the door. He handled it magnificently. The Commander tells me that for greater security he is going to send down air bells to serve as way-stations. Same principle as the house: inverted barrels weighted with lead and half-filled with air. If we run out we can swim over to one and put our head in the barrel, where there's enough air for five or six minutes. That will be most useful once oceanauts start using future underwater houses without surface support.

"There is a whole world here that only Claude and I know. This is the first time in 20 years of diving that I really have *time* to look around. The seaweed, for example, is fantastic, especially at night under floodlamps: it's swarming with seahorses, anemones, shrimp, and fish laying eggs. It's like attending the birth of fish. And then there are the ones that follow us around, always the same ones, almost as if we had arranged meetings. If we put them in an aquarium ten times as big as the one we built in Monaco, I'm sure they would come up and eat out of our hands.

"The Commander came down to have lunch with us. He brought along some green caviar. It's the pressure that makes it green, he said with a laugh. I don't know about that, but when he tried to uncork the bottle, the cork stayed jammed in due to the pressure. The damping of sounds, too. We ask the Commander to try whistling, but he couldn't. After practicing, Claude and I can now.

"The Commander is planning future houses at greater and greater depths, several houses at different levels. A sort of inverted Himalaya with base camps: camp I, camp II, etc., and in the deepest ones divers would breathe complicated mixtures of nitrogen-free gas. It would be tempting to stay on the bottom of the ocean. At Grand Congloué we worked several years at a depth of 130 feet. But after 15 minutes a shotgun was fired three times to remind us to surface. If only we had had the house then!

Conshelf II: assembling the underwater houses at Port Sudan.

Tugging the deep house from Port Sudan to Shab Rumi.

Divers preparing the platform for the underwater village of Conshelf II.

"The Commander is euphoric. Is it the wine or the pressure? He talks about colonizing this new world: men could live underwater with their wives and children; there would be schools and cafés—a real Far West."

Founding a village

The complete success of the Frioul experiment encouraged Commander Cousteau to attempt something much more important: creating not only a single house, but almost an underwater village. To make it even more interesting, it was decided to put it in the middle of sharks and coral: in the Red Sea. This was to be Conshelf II. A film would be shot showing the life of the oceanauts in a tropical sea setting. The film was eventually called *World Without Sun*.

The four underwater steel houses that were to constitute the underwater village were prefabricated in Nice and Monaco: the saucer garage, called "the urchin," a sort of service station where the diving saucer could park and refuel, the large house called "Starfish" that the oceanauts would live in for a month, the tool shed and the little house, or "deep station," where two divers could stay for one week. The "deep station" would be at 75 feet and the other installations at 30 feet.

These prefabricated houses were dismantled and transported from Nice to the Red Sea in an Italian cargo ship, the *Rosaldo*. They were rebuilt on shore and pulled to the spot where they were submerged.

The spot on the bottom that best fulfilled the criteria for the village remained to be found. The *Calypso,* which had left Nice on February 27, 1963, arrived at Port Sudan on March 16 after carrying out several biological research missions despite constant bad weather.

Here are excerpts from Falco's diary about that voyage:

March 8. "We enter the Straits of Jubayl. It's still winter. The sky is covered with clouds. At 3:30 P.M. we approach the reefs that surround the wreck of the *Thistlegorn.* With the depth gauge the Commander finds some large submarine cliffs from 180 to 90 feet down. I'm anxious to dive with the saucer. But the wind is blowing and it is already late, so the dive is put off until tomorrow. We drop anchor."

March 9. "The wind is blowing incredibly hard this morning. It raises clouds of sand on the coast, a real storm. The Commander got up at 6 A.M. to direct the saucer's dive, but he lets everyone sleep. The *Calypso* is pulling hard on its chain."

March 11. "The weather improves, the sea is calm. The *Calypso* sounds the bottom."

March 16. "At 7:30 A.M. the sea is so heavy that the *Calypso* is unable to turn at the lighthouse to enter Port Sudan. The Commander has to tack for over two hours before reaching the shelter of the reefs and entering the port. The pilot, an Englishman wearing white gloves, takes us to a mooring buoy, there being no berths available for the moment."

Exploration

Port Sudan had been chosen as logistics base for the Conshelf II experiment. It was the only port in the area equipped well enough.

Still, it was necessary to find a plateau 30 feet down large enough for the large house and the saucer garage. And a second story 75 feet down for the second

A triggerfish protecting its eggs.

house also had to be nearby. The area also had to be rich in fish and colorful coral.

At the end of March 1963 Commander Cousteau flew to France to supervise the finishing touches on the prefabricated houses that would be the homes of the first men to live for a month beneath the sea.

The *Calypso* remains in the Red Sea, where it has several missions to carry out. As he left the ship, Commander Cousteau told Falco, "Find me a good spot for my underwater village."

Falco finally found that spot after searching through the Red Sea for another week. He had to make three or four dives in the saucer per day, meaning daily stays of 5, 6, or 7 hours, an exhausting pace for his nerves. Aided by Coll and Sumian, he compared all possible sites and drew up lists of the advantages and disadvantages of each: the bottoms were too shallow or hollowed out, the corals were mediocre or the spot was too far from a sheltering cove where the *Rosaldo* and the *Calypso* could go in bad weather. Falco finally chose the reef and the lagoon of Shab Rumi, which appeared to fulfill all the topological criteria and was not too far from Port Sudan where the tons of steel houses would be unloaded.

Double page following: The underwater village of Shab Rumi. Falco works at setting up the saucer's garage.

Shab Rumi

Falco has recounted how the ideal site was at last discovered:

March 30, 1963. "This morning on the bridge there are dozens of grasshoppers that were attracted by the *Calypso*'s lights. A little cricket in the gangway sang all night.

"Large clouds are coming from the north, and the wind is beginning to blow again. We sail slowly around the reef taking soundings, looking for a plateau. But the cliff is the same all over, and the sounder indicates that the plateau is at 130 to 165 feet. Then the cliff descends to 900 feet. We can find bottoms at 120, 150, 300, and 900 feet down, but nothing at 30 and 75 feet.

"We moor the *Calypso* and I leave with a diving crew to explore the first pass. A short way past the entrance there is a long, sandy plateau framed by two large clumps of living corals. There are numerous fish of different species.

"After the ledge at 45 feet, the cliff descends to 120 feet, where there is another sandy ledge surrounded by a dozen coral bushes. Toward 120 to 150 feet I find a last ledge planted with a whole forest of coral. Beyond that the depths disappear into the dark. That abyss of 300, 600, or 900 feet is worrysome. If we put houses up there and one of them topples over it will be a catastrophe. We will have to keep that in mind by mooring the houses securely to the reef with large cables.

"It is the sort of cliff often associated with coral, having successive balconies and little loggias. It occured to me that the small house at 75 feet might simply float at its level, moored to a large immersed weight.

"The Commander will have to make the decision because this is the nicest area we have yet found. Also, the lagoon forms a harbor for the ships, being 140 feet deep in the middle.

"In the afternoon we dive into the second pass. There the cliff descends directly from 45 to 190 feet, and we go inside for a look around. As at the other pass, we find the bottom of the lagoon at 90 feet. Above us fish are swarming. Before getting back into the launch we cross a school of radiofish that come and play around us."

This ideal spot that was to become the first underwater village had a name on the map: Shab Rumi. It did not look like much on the surface: a reef with barely visible corals. In the center there was a pile of cemented stones surmounted by a beacon. Commander Cousteau visited the spot after a long talk in the *Calypso's* wardroom.

"This morning the spot will be chosen," Falco writes. "The *Calypso* is moored opposite the west pass, and the Commander glides along the keel with me 10 feet behind. The water is cloudy in the lagoon, and we encounter a current in the pass. We find the buoys that mark the second spot. We note that the water is cloudy only in the middle of the pass, whereas to the right it is clear. That is doubtless due to the current. We dive to the ledge at 120 feet and come up along the forest of black coral, returning to the first spot. The Commander signals to us with his finger: this is it. I agree, having thought this was the ideal place since the first day. Captain Alinat and Frédéric Dumas check the depth with a tape measure: it's 36 feet. I can see that the Commander is pleased. Leaving Captain Alinat to take the last measurements, we go over to the other side of the pass. The water is clear, and it's obvious that this is the right place. We return to the ship.

''Bébert, my friend,'' Cousteau says, ''you'll have to dive here again this afternoon with Alinat and Alexis Sivirine to set the stakes out for the first house. This is where we are going to put them, whatever happens.''

The hardest work

Plenty would happen. The hardest job would not be to live for a month in the sea, but to set up everything necessary for it. The crew never had worked so hard.

The plateau 30 feet down that would serve as the foundation for the large house was not absolutely flat. Actually it was a coral slope open toward the sea. It had to be leveled with picks and shovels. The whole crew worked at it until a better way was found in the form of a ''plow wagon'' hauled down from the *Calypso* that a diver guided with two handles in a cloud of sand and crushed coral. It worked very well.

The underwater construction workers used a pulley operated by the *Calypso*'s winch. On signal, the winch was set in motion, pulling the plow along the bottom as a diver guided it, shoving hundreds of pounds of sand toward the edge of the cliff with each passage. Large outcroppings of coral then had to be cleared away by crushing them.

The whole layout had been ''surveyed'' with little yellow buoys.

When the plateau was nearly flat, the crew had to go back to Port Sudan to put the houses together. The work was done on the docks in such torrid heat that the metal burned their hands.

The houses were made watertight so they could be floated.

"With Bonnici's help," Falco says, "I put four tons of lead weights in the lower part of the house while Maurice relayed them to us with a crane. The work was simple and fast. Four weights in a sling are brought to the door, then pulled inside to the middle of the floor. We walk heavily even though we have removed our flippers, piling the weights in containers."

Tugging

The house is immersed in the green water of the port. Then the *Calypso* tugs it slowly, at 2 or 3 knots, from Port Sudan to Shab Rumi, 35 miles away. The trip is made difficult by reefs and coral, and is not without incident:

"At 6 P.M. the houses's portholes leak and its inside pressure drops. The *Calypso* stops. I go down to plug in an air tube and note that the water is full of particles. I feel stings on my legs and then see that the particles are the stinging hairs of jellyfish. I'm able to avoid them on the way back up, but the stings I already have burn unpleasantly."

The *Calypso* sails on the 13th and arrives on the 14th, thanks to a calm sea all night.

It is impossible to immerse the houses immediately, since a lot of preparation of the site has to be done. They have to be moored at the spot in case of a storm.

Meanwhile, the *Rosaldo,* which served as logistics base for Conshelf II, was preparing to enter the lagoon of Shab Rumi where three anchors weighing 1100 pounds each have been sunk together with enormous cables.

The pass leading into the lagoon was narrow, and Falco had carefully noted all its characteristics. He had given a detailed map of it to the skipper of the *Rosaldo,* but at the last minute the captain noticed a strong current and did the opposite of what Falco advised: he approached the pass on the left rather than on the right. The ship was turning broadside in the pass and the captain had to put the helm hard over to starboard. He was lucky: the coral outcropping that could have torn open the *Rosaldo*'s hull only scraped it.

In the water the divers watch this mad maneuver anxiously. When it is over they get the ship moored inside the lagoon, behind the reef.

Commander Cousteau supervises the lowering of lead weights into the water.

A diver guides the weights toward the saucer's garage.

A diver walks on the bottom carrying a 100-pound lead weight.

Immersing the houses

To place the floating houses on the bottom, it was necessary to position them above their respective locations and then load them with lead weights. The work took several whole days and nights.

The large house was the most difficult. Launches carried the weights from the *Rosaldo* to the site. The crew redoubled its effort, fearing that if the sea became heavy the houses would sink pell-mell and the whole operation would be compromised.

Falco's notes mention the following:

"At 9 A.M. on June 6, 1963, Captain Perrier, who was handling the assembly of the saucer garage, has 400 pounds of sheet metal fall on two fingers of his left hand. The doctor operates and Perrier has to leave for France three days later."

The divers placed the weights one by one through a panel in the roof. It took several tons of them to make it sink. When it was about to go down, the door and panel were quickly closed, other weights were added, and it was sent to the bottom. To stabilize it there, tons of other weights were placed on the plaques to which the house's telescopic legs were attached.

In all, the crew handled 4,790 weights of 100 pounds each. Occasionally the divers had to move them several times. The garage for the saucer sank too fast for lack of enough air being pumped into it. It hit the bottom so hard that the bolts broke on its telescopic legs and it fell over. The crew had to remove 20 or 30 tons of weights, float the garage, repair its legs, put bolts in and sink it anew, this time adding air from time to time to compensate for the weight of the lead. The weather worsened during the night, and the next morning it was the large house that had half-sunk.

"We had to start all over several times," says Falco. "It was very hard work, and the crew was starting to get tired. I was sick of it myself, and began to wonder if we would ever get the whole job done. Twice the "little house" supposed to go to 90 feet sank to 150 feet. People who see the film *World Without Sun* can't imagine what we went through to get that village going. Some of the incidents happened when the houses were already equipped. And God knows the equipment was sophisticated: electric lighting, heating, telephone, kitchen, air-conditioning. Each time we had to take the motors out, change the wiring, and dry it all out."

As to the little house that was supposed to be set up at 75 feet, the *Calypso* lowered it with the winch and it was floating at the end of a cable when the current made it start turning. The house broke the cable and sank to 140 feet, where divers had to reattach it to the winch, pump it with air, and hoist it back up to its site.

Unfortunately its electric equipment had already been installed. It had to be emptied and the motors removed to the ship, where they were dried in pressure cookers. The pressure cooker proved to be an indispensable tool all during Conshelf II.

"All this took about one month," Falco explains. "When everything was in place, when the compressors and motors were working, the experiment proper could commence. We were very tired. Several of the men fainted and Christian Bonnici was sick. I was trembling with fever all during the construction. We were in the water from 6 in the morning until 8 at night."

A friendly triggerfish

"When I had discovered that platform there were large sharks around it. During our first dive to it we had to turn back and climb up on the reef because two of them came too close for comfort. But as soon as we began putting houses under the water we saw no more sharks."

When the village was ready, only once did two small sharks approach the houses. But all the other fish that slept at night in the pass continued to live there normally. The divers even made friends with them, especially the triggerfish, who became extremely friendly and followed the men during their dives. They knew they would be fed. The divers opened giant clams for them, their favorite food.

These friendly triggerfish could also be an inconvenience. They sometimes bit at the divers' ears because it looked like clam flesh to them. They have formidable teeth, with which they can break an oyster shell. They can lift clumps of coral with their jaws to look for crabs and shellfish.

A large triggerfish always accompanied the divers when they went from the large house to the one at 75 feet. It especially liked to rub against their vinyl suits. That one became the favorite of Guilbert, one of the oceanauts. When Guilbert tapped on the porthole from inside the large house, the triggerfish came and looked at the men, rolling its big eyes and opening its small mouth as if it were trying to make bubbles. But it was another fish, Jojo the Grouper, that became famous in *World of Silence.* Many other fish lived permanently in the village. They showed no fear and swam around excitedly at mealtime.

The deep house is taken down to 75 feet.

The site was dirty with broken coral and all the clouds of sand that had been raised. The water was cloudy, and to make good motion pictures the area had to be cleaned. The suction pump did not help, and finally the *Calypso* stopped just above the village and started its screws, blowing the sand and debris away.

Falco's diary for June 13:

''We spend the day on finishing touches and cleaning up the site. . . . The electricity creates the most problems. The pressure drives water into the cables providing current to the house at 75 feet, and on the surface bubbles and drops of water cover the wiring.

''In general the village is ready and waiting for its inhabitants. Dumas will be the first; he will spend the night in the large house to give his opinion on its habitability.''

Exit lock of the main house.

Cinema

As soon as the house was finished and inhabited, filming began to record this new underwater life.

Eight divers wearing silvery suits went down and swam around a large coral bush before going to the city. This was the first sequence of the *World Without Sun*. Shooting it was tricky. In order to give the impression that the divers fell from the sky into the sea, they were suspended above the water by their legs, attached to a square frame held by the crane. At a signal from a shotgun they all dived at the same time. Hanging from that frame by their knees was no small matter for men wearing their air bottles and lead belts.

Bringing them out of the water was just as athletic. Falco and his colleagues surged through the surface holding onto a rope, an arduous stunt after spending over an hour under the water. But the most unpleasant part was wearing isothermic clothing in suffocating heat.

On June 15, five divers waved good-bye and plunged into the water to live in the sea. They were Professor Vaissière, Wesly, Folco, Vannoni, and Guilbert. They leave the sun behind for a long time.

The next day Falco and two divers brought the oceanauts their most precious equipment in watertight caissons: microscopes, a typewriter, laboratory glassware.

Cousteau later went down to live in the large house for a day or two. Falco writes:

"I stay on the *Rosaldo* with the film crew, and after dinner we go shoot a night scene of two divers petting a parrotfish paralyzed by our floodlamps.

"Dumas and I go say hello to the oceanauts, who have turned on all their flashing colored lights and the revolving beacon on the large house. Behind the picture window I can see the Commander smiling. He laughs when he sees me in the water. He looks really happy.

"In the garage for the scooters, Dumas throws his light on a big grouper and I pet it going by. It doesn't seem to like that, and bumps against a bulkhead getting out."

June 17. "I will help film scenes inside the large house. With the floodlamps turned on the heat is unbearable. There are 12 of us in there, all sweating heavily. As soon as they don't need me any more I run for the water."

June 24. "Bonnici and I have lunch in the underwater house. The air-conditioning is working well and it's cool. It is very pleasant to eat in front of these two picture windows giving a view of the bottom of the sea.

"The triggerfish tamed by Guilbert, the cook, becomes a nuisance, waiting beside the exit ladder to bite at divers' ears to get something to eat. Inside the large house there is a parrot that was brought down in a pressure cooker. It's supposed to indicate by its behavior whether the breathing mixture is dangerous. It is talking and shrieking a lot, and certainly seems to be breathing well."

A diver, accompanied by a surgeonfish, looks through the window of the large house.

The wreck of a house

June 26. "Aboard the *Calypso* at 11 A.M. a voice announces that the small house has sunk. I run to the afterdeck and see Roux on the ladder. He shouts, 'Servelot is still in it.'

"A minute later I've got my gear on and am in the water. The house is jammed against the cliff face. I just manage to slip between the grill and the coral, but when I push against the door it's blocked. I swim around to the upper porthole, but the compression has fogged over the plexiglass and I can see nothing. I return to the door, and this time it opens half-way when I push with all my strength. I feel around it, afraid I'm going to find Jean Servelot's body, but all I touch is a wooden box that I push out of the way. I go up to the second chamber and find Servelot, who tells me calmly that everything is all right. I breathe a little easier and my heart slows down. I give Servelot a scuba, and, finding neither his mask nor his flippers, I tell him to relax and let me guide him. Christian Bonnici is waiting for us at the exit. He helps me pull Servelot against the current. As I think about it, I become convinced that the house was overturned by the current, which must have made it lean over and then created a suction effect in the entrance. An hour later the house is floating again, helped up by the *Calypso;* the electricians will have to start all over."

The stay at the bottom of the sea was not always full of drama. The film was shot without any problems, and the health of the oceanauts, closely watched by a doctor and a physiologist, was perfectly normal after three weeks. Their morale was also good. To be sure, there was plenty to keep them busy. They left their domiciles every day to photograph or capture fish for the Monaco Oceanographic Museum.

Canoë and André Portelatine, the occupants of the small house—which was finally stable—stood their sojourn at 75 feet very well. The mixture of gases they breathed—oxygen-helium—gave them a strange, fluty voice: when they phoned the large house, everyone always laughed about those funny sounds.

But fatigue was noticeable after all that effort to set the village up and make the film. Canoë came up too fast after a long dive, got the bends, and had to spend six hours in the Galeazzi decompression caisson. Many divers were bothered by their ears, and Bonnici developed a detached eardrum jumping from the launch. As for Falco, he had a pain in the stomach from running into a ballan wrasse weighing 20 pounds that was trying to get away.

On July 23, the experiment could be considered to be over and Commander Cousteau gave orders to begin dismantling the village. The heavy work started all over as the lead weights had to be pulled out.

The large house was lightened and floated to the surface. All that remained inside was the parrot, which had never talked so much. Soon it would be set free.

"When the sand settled," Falco writes, "it was a desolate sight. Nothing remained of our village but bits of debris. Happily, the coral will soon cover all that over.''

Falco was right. The sea began to erase all trace of the men who had lived in the heart of the waters for so long. The *Calypso* returned to Shab Rumi four years later: on the cables, iron bars, and ''fish houses'' left behind, numerous sprigs of coral had grown. They were already as big as a fist. They have certainly grown since. The spot has been transfigured. Gorgonia, madrepores, and Alcyonaria are burying what is left of the village beneath the sea.

8

Conshelf III, or life at 300 feet down

A TAILOR-MADE PLATFORM—A STEEL SPHERE—SETTING UP —300 FEET DOWN—THE ELEVATOR—DRAMA IN A STORM —A RECALCITRANT HEAD—WORKING—MISSION ACCOMPLISHED —RETURN TO PORT

The experiment at Shab Rumi had shown that five men could live for a month at a depth of 30 feet and two others at 75 feet for a week without impairing their health in the least. It had proved that man could stand a stay at the bottom of the sea with his blood saturated with nitrogen. But Commander Cousteau wanted to mount an even more important operation: he wanted to demonstrate that six men could live for one month at a depth of 300 feet and could work daily with their hands on, for example, a wellhead. These were the criteria for the experiment named Conshelf III, or E.P. III. It seemed unnecessary to situate it in far-off tropical seas. It was decided to stage it in the Mediterranean for greater convenience, if possible near the coast of Provence. Once again Falco was charged with finding the ideal spot.

During their training, the oceanauts practiced breathing from several emergency air bells in the area.

A tailor-made platform

He began his search off Cap Ferrat on August 24, 1965. He had to find a plateau 30 feet by 30 feet 300 feet below the surface and near another plateau 360 feet down. The wellhead would be placed on the latter.

By following the 300-foot isobath he finally found a small platform just the right size about 1,500 feet south-south-east of the Cap Ferrat lighthouse. To the west he found some cemented rocks, toward the north a slope of muddy sand covered with shells and a triangular stone inhabited by an octopus. Pushing farther on he found a slight sandy slope, and, toward the south, before reaching the edge of the greater depths, there was a small valley bordered by rocks covered with rare white and yellow sponges in the form of branches. Farther south, at about 90 feet from the platform, he noted a small trench ten feet wide. Beyond that a large slope descended toward the abyss.

To measure the area of usable terrain, Falco guided the saucer over it at a constant speed. André Laban noted the time carefully, and from that could be calculated at least the approximate dimensions.

A steel sphere

The underwater house of the Conshelf III experiment was composed essentially of a habitable sphere attached to a chassis that could float at various levels and be stable on the bottom. It weighed 130 tons. It was made of steel, and it included an entrance lock and an upper panel for loading heavy equipment at sea or on land. It had three portholes.

Inside, it was divided into two stories: the upper one was a living room, around which were arranged the laboratory, the electric control panel, the kitchen, and the cooling system. On the lower floor a partition separated the six-bunk dormitory from the "wet room," where diving equipment was stored and the toilets and showers were grouped.

Ballast tanks holding air or water gave the structure a floating capacity ranging from +15 to −30 tons. Besides these, containers holding 40 tons of iron pieces and 32 tons of granulated filings provided variable security ballast.

For the first time in history, this underwater station was to be linked up to a permanent information center incorporating an IBM 1620 computer. The computer would give instant information on the composition of atoms in the gas mixture breathed by the house's occupants.

The house was to be filled with 150 bottles of helium containing 12 cubic yards each. Thirty bottles of reserve helium were stocked in the sphere.

It was in the port of Monaco that the new oceanauts took their places in the big floating ball painted with a yellow and black checker pattern. There were six of them: André Laban, head of mission and director of the French Office of Underwater Research, Christian Bonnici, Raymond Coll, and Yves Omer; three experienced divers specially trained in wellhead techniques; Philippe Cousteau, the cinematographer; and Jacques Rollet, a physicist from the Monaco Oceanographic Museum.

Albert Falco had the important job of watching over the safety of the oceanauts for the duration of the experiment. If the sphere's occupants ever had to evacuate it because of fire or an accident in the gas mixture, they could get into two Galeazzi diving bells situated on each side of the house. Falco would then go down immediately in a scuba to insure that the bells were closed and to cast off the moorings that held them to the sphere.

Falco also had an important role to play during the positioning of the "ball" on the bottom. Off Cap Ferrat, aboard the diving saucer, he had to guide the surface maneuvers by telephone during immersion of the occupied sphere.

Setting up

On September 17, in the port of Monaco, the six divers enter the water without a scuba, swim to the entrance lock, and disappear into the sphere. This is child's play for them because for a year they have undergone strenuous training. They have dived with a helium-oxygen mixture to 330 feet. During the previous two weeks they went through every drill that can be practiced underwater: removing a scuba and swimming 100 feet to take someone else's mouthpiece; going to an emergency way-station, going back, and finding the scuba equipment; swimming without a mask; sharing an oxygen bottle; and leaving the scuba on the bottom before going up to the surface 90 feet above.

At 1:30 P.M. on September 21, Commander Cousteau announces over the radio to the sphere's occupants that the *Espadon* has just entered the port, aided by the trawler *Lutin*. At 4:45 the next morning, the convoy is in sight of its destination as a brilliant sun rises on the horizon.

Then comes the tugboat *Physalie* from the Cap Ferrat lighthouse pulling a long bundle of electric cables looking like a huge sea snake. The cables float on the surface, buoyed by yellow 50-gallon barrels every ten yards. The cables will be connected to the underwater house, bringing light, heat, and air-conditioning to the six men who will live there. Of the electric equipment to be powered, the most precious are the compressor and decompressor that will enable them to dive daily to depths of over 300 feet.

300 feet down

Immersion of the house is all the more difficult because the plateau is just large enough to accommodate the sphere's four legs. Falco in the diving saucer guides the crew on the surface by telephone.

At first the ball is held back by inertia and refuses to sink, so it is loaded with more weights. When the descent begins, it starts to accelerate dangerously and the divers barely manage to slow it down.

The Conshelf III experiment begins: the oceanauts—Raymond Coll, Philippe Cousteau, and Christian Bonnici—in the port at Monaco. Above them is Armand Davso.

Inside the underwater house Raymond Coll, Philippe Cousteau, and Christian Bonnici prepare to leave for a work session.

At thirty minutes past midnight on September 22 the sphere, still supported by the big winch on the landing stage *Labor* is in sight of the bottom. In the saucer, Falco gives instructions to the surface, especially to the *Physalie,* which is bringing in parts of the house from the northwest to the southwest.

At last the articulated legs touch the plateau and the house settles there with a slight list toward the coast.

At 1:30 A.M. Falco casts off the saucer's ballast and rises slowly past the imposing black and yellow sphere enveloped in the dark water.

The next day at 6:15 A.M. Falco lands the saucer to the east of the house. Knowing his way around the sphere, he goes to the entrance lock where a gleam of light illuminates the water. Flippers appear and descend the ladder awkwardly, entering the water in a burst of bubbles. There is the outline of one diver, then a second. They both head east where the cable that carries the food containers is located.

Stretched between a buoy and a dead man on the bottom, this cable is the track for watertight caissons that carry equipment and food. This sophisticated elevator replaces the picturesque pressure cooker used during previous experiments.

Thus Falco sees, as he will every morning, that a new day is beginning normally for the oceanauts and that routine is being respected. This liaison between the surveillance ships and the bottom is established every day at dawn.

The elevator

Sending down the food container is not always easy for the surface crew. Canoë Kientzy and Bernard Delemotte, who manage supplies, are hard put to fill the caissons with the *Espadon* rolling heavy under a strong wind from south-south-west. The heavy container goes overboard, and Falco and Delemotte lend a hand to the divers struggling to get it attached to the cable. After several tries they get it snapped on, and Bernard opens the air valve on the ballast tank. A long string of bubbles shoot out and the container sinks toward the seven-ton dead man. The divers can tell when it reaches bottom by putting an ear against the cable to listen to the pulley or simply by feeling its vibrations with a hand.

A diver works 375 feet down on the "Christmas tree" illuminated by the diving saucer's lights.

On the bottom, two oceanauts then unhook the container, which stirred up a cloud of mud when it landed. They carry the heavy box to the entrance, attach it to a pulley, and crank it up into the sphere.

Then they bring down the previous day's container, either empty or filled with waste, hook it to the cable, and send it up.

On September 24, 1965, the sphere's decompressor breaks down, just when the oceanauts are supposed to figure in a film. That shooting schedule has to be cancelled, but Falco takes two cinematographers down in the saucer anyway. Georges Alépée and Michael Deloire both get some shots of the oceanauts swimming back and forth to the elevator to get the spare parts to repair the decompressor.

Drama in the storm

At dawn on September 27 heavy swells from the south-south-west sweep away the platform built on the rock at the foot of the Cap Ferrat lighthouse. The platform supported the electric cables, which now are rubbing against the rocks and are in danger of breaking at any moment.

Commander Cousteau at the headquarters on Cap Ferrat immediately grasps the seriousness of the situation and asks the oceanauts to prepare for surfacing. Meanwhile, Falco and several sailors from the *Winaretta Singer* go to the spot where the platform was, but are battered by huge waves and one sailor is nearly swept away. Falco nonetheless is able to get an arm around the pole that carried the cables and begins to pull it up to get them above the water. He battles with the waves that break over him in clouds of foam, realizing that the sphere's electrical supply can mean life or death for its occupants.

The sailors from the *Winaretta Singer* pass nylon cords to Falco, who ties them to two supporting pulleys and gets them solidly anchored to points on shore. Headquarters alerts the sphere's occupants that everything is in order and that they are out of danger.

In the afternoon of the following day the sea calms and helium bottles can be sent down the elevator to the oceanauts. Commander Cousteau decides to set up

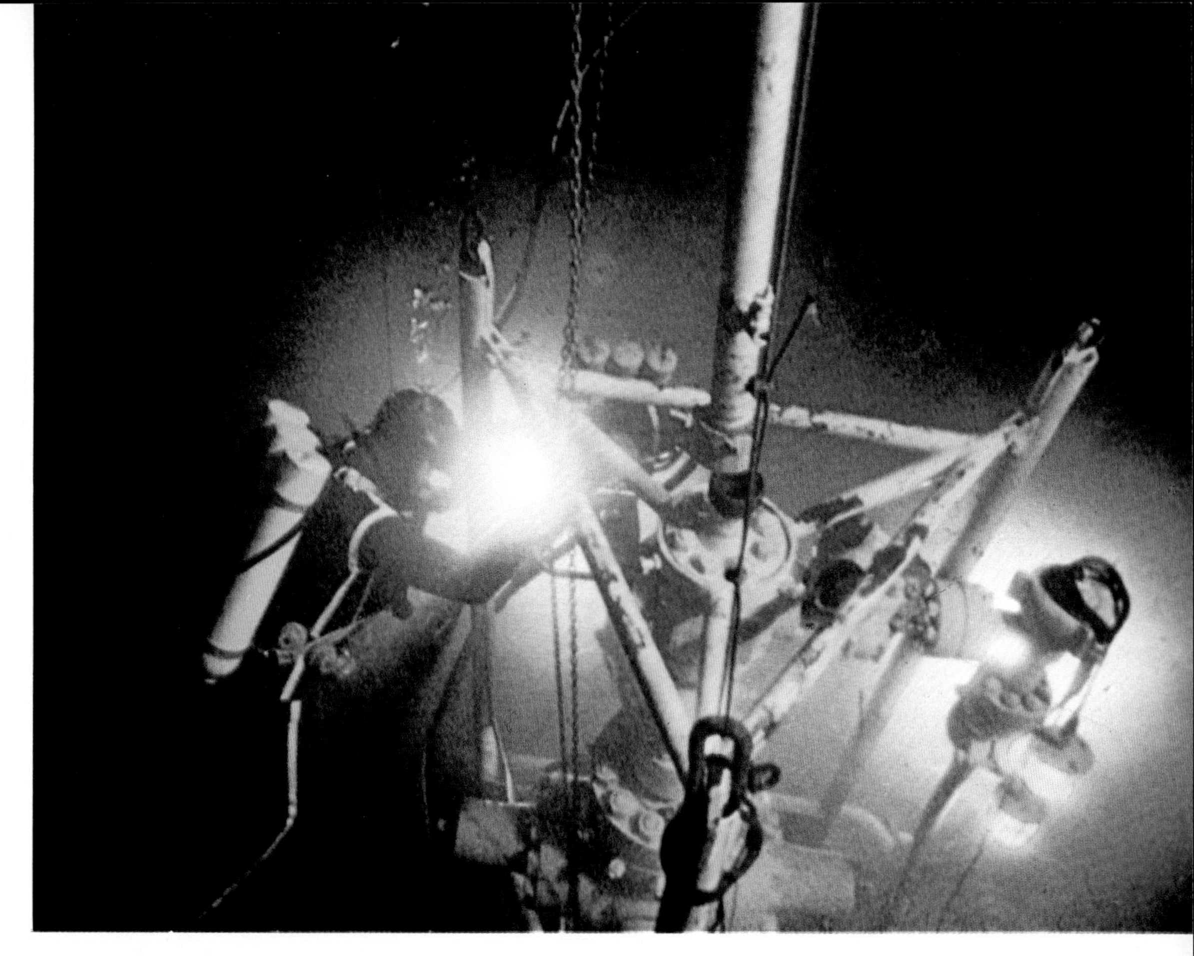

Divers working on the simulated wellhead.

the wellhead which, at 387 feet, will demonstrate the ability of divers to work at great depths. The maneuver is not going to be easy, with the sea still rough.

A recalcitrant head

Nonetheless Yves Bousquet and Albert Falco go to the site in the saucer. What is known in oil exploration as a ''Christmas tree'' will be located on a small ledge. The saucer will coordinate installation via a telephone line.

"After a few minutes," Falco recounts, "we see the wellhead, periodically disappearing in great clouds of mud. I'm afraid it's going to bang into the saucer. Our telephone line gets caught in the rocks, and I ask the surface to send the oceanauts over to free it. Twice they have to suit up and work in the icy water to untangle the line.

"After several tries, I get the wellhead on the ledge, a foot from the edge. Yves Bousquet and I had several scares that the wildly swinging thing was going to crash right into the saucer.

"Bit by bit things calm down. The mud sinks back to the bottom and the lobsters go back to their holes. It was 9:30 P.M., and we had passed over three hours on the bottom to position that Christmas tree for the oceanauts."

Working

October 12, 1965, is the 12th day that Conshelf III is on the bottom. The weather is good, the sea calm, and Falco takes an American writer, James Dugan, aboard the saucer.

Turning slowly, the miniature sub lands in the mud to the east of the sphere. Falco drops off the weight and lands using internal ballast. Seventy-five feet away he can see the imposing mass of the house. He arrives just in time to see the oceanauts exit and get tangled in the air tubes. They have to pull around not only their air tubes, but also the electric cables on their lamps, which would be cumbersome on land, much less in the ocean. Then they reach the slope and head for the wellhead.

Their first job is to install a "buoy sphere," containing about 200 cubic inches of air, to support a pulley that will help lift heavy equipment. The saucer approaches them slowly, and Falco recognizes Yves Omer and Christian Bonnici through their masks. Their movements are slow but precise as they rise and descend and signal to each other.

"I feel happy and proud," Falco says, "to see them so at ease because I helped train them. This is my reward. Of course, I would have liked to be with them, but at least I'm providing light for them with the saucer.

"Not far off I can see an enormous anglerfish waving its antenna, trying to attract fish for its meal. Farther along in the blue light a lazy moonfish swims by awkwardly, reflecting light from our floodlamps."

Gradually the icy water gets to the divers despite their triple-thick neoprene suits, acting more intensely because they are breathing helium. They begin to tremble.

Still, they hang on, and an hour later the ball hangs there in the water like a giant jellyfish. Then they attach the pulley and soon they will be able to begin disassembling the wellhead. The work program called for them to plug the well under pressure in order to take the valve off for cleansing or repairing. This necessitates lifting the air lock tube, weighing 440 pounds, which accounts for setting up the floater with the pulley. But cold begins to paralyze the divers, whose movements become uncoordinated, so they leave for Conshelf III to warm up.

On October 4 and 5 the oceanauts devote all their time to the wellhead, except for sessions of operating the elevator to get their food, which sometimes took rather long.

"All week," Falco says, "I dive with the saucer two or three times a day and take part increasingly in their underwater life. I can tell from their looks that I furnish them moral support with the saucer as well as assistance in the form of illumination."

Mission accomplished

At about 12:30 P.M. Commander Cousteau slides through the saucer's narrow hatch. This is the 19th day of the experiment, the 14th day that Conshelf III is on the bottom. The oceanauts feel more and more at their ease, and they work without losing a second. Their work schedule calls for both hard labor and precision in removing and replacing the plug and its paraphernalia and in changing the packing of the master valve, and so forth.

Christian Bonnici, as chief diver of this operation, sets new records for staying underwater. He spends more than seven hours in the water out of 24. He has to get the device in perfect shape to stand a watertightness test of 1,700 pounds per square inch.

You can almost say that the work is easier for the divers than it would have been on the surface because they could easily move up and down the 45-foot structure and work in awkward positions more comfortably.

The operation is repeated several times, and at the end it takes less than one hour, which is about the same as on land.

On October 5, after a four-hour dive, Cousteau and Falco surface, having used up all the rolls of film they had.

From the 5th to the 9th of October, they go down every day and spend over 30 hours around the oceanauts. Each time, they signal their arrival by a flash of floodlights through the house's portholes.

"Every time I see the oceanauts," says Falco, "I can't take my eyes off those elegant silhouettes dressed in black with a luminous yellow stripe down the side. They are really masters of the depths: they move easily in all dimensions with their double air tube, unhindered by gravity. This underwater choreography reminds me of the astronauts' walking in space outside of their orbiting capsule, all the more so since the thousands of plankton and tiny jellyfish picked out by our floodlamps seem to drift toward infinity."

On October 9 the oceanauts successfully finish their work on the wellhead. The whole operation has been followed on closed-circuit television in the operational headquarters at Cap Ferrat.

The inhabitants of Conshelf III then begin their scientific experiments. Jacques Rollet, a physicist, is charged with gathering samples of sediment for the Monaco Oceanographic Institute. Then he makes studies of micro-currents using calibrated balls that fall into a target area. The third important experiment is devoted to artificial photosynthesis, and is directed from the surface by Professor Jean Brouardel.

The divers could work in different positions and at all levels of the wellhead.

Return to port

On Tuesday, October 12, the oceanauts stay in their diving suits nearly all day. The diving saucer with Albert Falco at the controls stays down with them. Commander Cousteau is with him, filming the disassembly of the Christmas tree's electrical system. It is then that Falco notices that Yves Omer is in trouble: he is trying to climb up a steep slope carrying a bunch of cables. The cables get caught on the bottom and Omer stumbles with each step, for the load is too great for him to swim.

"I want to help him," Falco says, "and I get the idea of pushing him gently with the saucer while lighting his way to the house.

"Once he gets there all right, though trembling with cold, Omer thanks me with a wink and a wave, and leaves to help the others who are trying to get together the equipment dispersed around the house. For them it is the last time they will swim around it.

"I go down one last time with Georges Alépée on October 13 to film the house as it surfaces. Besides helping with the film, I am supposed to let the surface know when the granulated filings begin to drop to lighten the house.

"Twenty minutes after we reach the spot the surface warns us to be on the lookout.

"Suddenly filings begin to spurt from the house and begin to form a cone three feet high. But the house doesn't budge. I tell the surface that its feet are still on the bottom. The Commander phones André Laban, chief of mission, to let air into the ballast tanks. At 5:25 P.M. the house begins to rise slowly toward the surface, its four legs sucking up a cloud of mud. I cast off my surfacing weight and follow the house up through my large porthole. But half-way up it accelerates and leaves my field of vision.

"At 5:32, to the general satisfaction, the house surfaces like a sperm whale. Divers disconnect the cables linking it to Cap Ferrat and plug in power from the *Espadon*. Then Conshelf III is tugged toward Monaco."

Two divers send down the watertight food caissons on the elevator.

On October 17 at 9:50 P.M. the oceanauts climb out of the sphere's upper door. The grandest experiment in survival ever attempted at great depths was over.

During a storm Falco climbs on the pole that supports electrical cables to the house and raises them several feet.

9

divers of the impossible

OVER 300 FEET—TRAINING WITH HELIUM—FIRST EXPERIMENTS
—IN CORSICA—SAINT ANDREW'S CROSS
—WITH THE DIVING BELL—SHOOTING THE FILM

Falco had begun his discovery of the sea's depths by apnea diving and, thanks to spearfishing, had acquired marvelous training: he was able to dive to 60 feet and stay there over two minutes.

The scuba enabled him to reach much greater depths and to stay there longer. It was a revelation and a liberation for him. Still, he did not make his first really deep dive, to 210 feet, until November 1952, when he tried to save Servanti at the foot of Grand Congloué. For the next ten years he was one of the best divers of the *Calypso,* if not the best, and he became the crew's chief diver. He dived innumerable times to 210 and even to 270 feet using compressed air.

Albert Falco beside a branch of coral 250 feet down in the Gulf of Bonifacio.

But he was not immune to the troubles that afflict all divers. He knew the sea well and could work efficiently at great depth without becoming exhausted, but at certain times he still felt the narcosis, or stupor, that can affect divers.

"Your physical condition can considerably influence your reactions in the water," he says. "If you have slept badly or been upset, that's enough to make you feel fatigued or nauseated, or even lose consciousness.

"At 120 or 150 feet down, the air you breathe through the regulator tastes like iron. Some divers compare it to copper. In any case, it is an unpleasant metallic taste. You sometimes also have a ringing in the ears. The further you descend, the more the brain is affected, and your ideas can become confused or strange. You have to leave the surface with a work schedule in mind and keep repeating everything you are going to do.

"The only way you can keep your heart from beating like a drum is to keep your movements slow and easy. Deep diving is as good a way to learn to keep calm as yoga. You must never try to push yourself. Some days I can't seem to find my breathing rhythm and my heart pounds like a trip-hammer. So I don't fight it. Sometimes you can ascend a few feet and feel better.

"At great depth I often feel uneasy until I have made several dives to the same spot and learned my way around it. That's why it's easier to work on a familiar site where you know the layout. There's no doubt that at some depths a mortal accident is always a possibility. The least error, clumsiness, or slowness in understanding the situation can cost a life. You have to talk to yourself constantly, repeating what remains to be done, what has already been done, what you must not forget. The hardest is to stay logical. Personally, I keep thinking of two things on the bottom: the work, but also my safety. 'Is your reserve bottle closed? Check the pressure gauge . . . your watch . . . how much time do you still have? If you run out of air are you ready to keep calm going back up? Is the safety belt easy to drop off?' "

Over 300 feet

This experience that Albert Falco had acquired over the years enabled him, on July 6, 1964, to dive to over 300 feet using only bottles of compressed air. He accomplished this exploit in the Strait of Messina in the company of an excellent diver from the *Calypso*, Raymond Coll.

"That day I crossed a new frontier," he says. "The day before I had gone down in the saucer with Mr. Guierman, a geologist from the Monaco Oceanographic Institute. He was making a geological study of the Strait of Messina to determine where the piles of a large bridge planned by the Italians should go.

"During that dive we had found terrifically hard rocks on the bottom: the saucer's claw could not break them.

"I suggested to Commander Cousteau that I go down with a scuba to get a rock sample with a hammer and chisel. I wanted to do it because I knew that we were going to dive with helium on the next cruise, and I wanted to see the difference between a dive to 300 feet with air and one with helium. Coll and I were in good shape.

"The Commander finds a rock at 210 feet with the sounder and marks it with a buoy. Our dive is planned for the next day at 11:30 A.M. to catch the slack tide because there is a violent current in the Strait of Messina.

"Coll and I descend along the nylon of the buoy in limpid blue water. At 90 feet we pass some jellyfish.

"At 210 feet I have a warning when I begin to see spots before my eyes. I feel a virtual explosion in my skull. I don't know what happens and I never will. I have been swimming with my head down to go faster. This stops me short and I am about to go back up in panic. I say to myself, 'I'll go up a few feet and think it over.' Then suddenly I feel better, I'm lucid and everything is normal. I look at the bottom and see it only 30 to 40 feet away. I hesitate a few more seconds and decide to go on.

"It is crazy, but in a state of stupor and wanting to go all the way down, I go deeper than 300 feet: I go to nearly 350. It's easy to go the last 40 feet to touch bottom. My pressure gauge is blocked at 300 feet.

"I signal to Coll, who also seems in good shape. I go over to the rocks covered with algae waving in a slight current.

"I take my chisel and tap on the rock, but it's impossible to break it. I sense that I am in an extreme state of stupor. I hit the rock once more and only a few algae come off. I can't break the rock itself.

"Now I understand that I am dangerously affected by narcosis. I've got to get back up. I signal to Raymond, who is a few feet above me.

"Nevertheless I would like to bring back something from the bottom. I grab some seaweed in passing and put it in the net bag that I tied to the nylon cord. I check my watch: 5 minutes 30 seconds have passed. I have to ascend a few feet and quick.

Recco shows Falco the best branches of coral he keeps in his shed.

The *Calypso*'s crane places the Galeazzi bell in the water.

"The nylon reaches to the depths. My heart flutters and I know that we must go up to calm it.

"I signal to Coll and move my flippers faster. He follows and we rapidly reach 120 feet.

"I head for the bubbles we have left behind us. I see Bernard Delemotte on the surface, and he shows us the buoy. We go toward it and begin our decompression pauses.

"Only then do I have time to think about our adventure, which, I must admit, was a little crazy.

"But for a long time I had wanted to know what my reaction would be at that wall of 300 feet.

"During our pauses we saw the long strings of seaweed I had taken out of the net bag floating in the blue. That was proof that we had gone to the bottom. Only the extraordinary hardness of the stone had kept us from bringing back samples of the rocks."

Training with helium

The crew's training with helium began two months later.

The *Calypso* had brought a Galeazzi diving bell from La Spezia that was to serve as decompression caisson during deep dives.

The divers were first subjected to training with air to accustom them to opening the bell's lower panel in the water. This panel can only be opened when it was in equi-pressure. Each occupant of the bell has to leave it holding his breath to pick up a large bottle of helium attached to the outside of the bell. The divers put on their gear in the water and practice each gesture until they are perfectly coordinated. They are training to operate around the Galeazzi without a scuba, which they have never done before. They descend the bell to the first decompression stage, about 120 feet, and then establish its equi-pressure. They then open the door and exit to exchange bottles or put them in place outside the bell, then return and close the door. If the door is not properly closed, it can lead to the death of all the occupants. It was owing to a badly closed door that there was the accident that killed two men during the H. Keller experiment. Thus there is is always the importance of practicing these maneuvers carefully.

The divers have to save as much time as possible during their maneuvers because decompression with helium is more touchy than with air.

The experiment requires using large bottles that are very heavy, and they have to be lightened with spherical floaters. That also means that at the end of the dive they pull on the divers' backs.

Then simulated dives start to 210 feet with divers breathing a mixture of helium and 16 percent oxygen for two minutes. This is a pleasant surprise for the divers: whereas with compressed air they felt slightly ill at 210 feet with the first symptoms of narcosis, when they changed to helium their mind is cleared even if it does taste at first like floor wax.

On September 13, 1964, Captain Alinat initiates the crew in preparing the mix of helium and oxygen: 260 pounds of helium to 46 of oxygen. Then he shows them how to check the mix using a Bekman analysis, and how to obtain a thorough mixture by shaking the bottles, first putting them in the water to facilitate handling their 100-pound weight.

First experiments

The first experiments at great depth take place to the west of Cap Ferrat.

It is noon when the *Calypso*'s crane puts the Galeazzi diving bell into the water. Captain Perrien rides with it to open the door at exactly 128 feet, while Christian Bonnici and Albert Falco harness their big bottles and, bending beneath the weight, go to the ladder on the *Calypso*'s stern.

When Captain Perrien telephones that he is coming up with his scuba, the divers synchronize their watches, fill their lungs, and plunge in.

They pick up speed fast, crossing Captain Perrien at about 90 feet, then passing the bell with a nylon cord dangling to the bottom that the divers follow.

"I look at my pressure gauge after two and a half minutes," Falco says, "and it says 250 feet. A reflex makes me stop and change position, continuing down *feet first,* which keeps the brain from becoming congested. But my mind re-

Double page following: Illuminated by the saucer's lights, Recco detaches a tuft of coral with his hammer at 300 feet.

mains exceptionally lucid. I check my depth again: 300 feet. It's incredible that I have not felt any discomfort. In fact it's so unusual that it almost scares me.

"The night around us is irrefutable proof of how deep we are.

"I come to the lead weight suspended 25 feet from the bottom. I undo the knot of the nylon leader and slide down it about 30 feet to the soft mud of the bottom, followed by Christian.

"I kneel on the bottom and look at Bonnici's suit, all wrinkled by the pressure. I ask with a gesture whether everything is all right. He gives me the O.K. sign and looks to be in great shape.

"We start the planned exercise: making knots on the nylon leader. Then I swim in an arc over the bottom. My bottles jiggle on my back, but that's no bother.

"I also do a few mental exercises. Everything is all right. I realize that eight minutes have already passed. I signal Christian and we return to the weight, then start up, hand over hand.

"At 180 feet, when we accelerate, I get vertigo. Everything starts turning around me. I close my eyes, and when I open them a few seconds later it's over.

"Fortunately I have already experienced this before and I know that it is due to a disequilibrium in one ear.

"Thirteen minutes after leaving the surface I unharness my bottles and attach them to the bell. Slowly, in apnea, I enter the big steel cylinder through the lock and surface inside. When I talk over the telephone I have a voice like a duck's, the usual effect of helium. Christian Bonnici enters the Galeazzi right after me. We close the door and signal that surfacing can begin.

"We make the change from helium to air in the bell without difficulty. There's no problem during the decompression, which lasts one hour forty minutes."

Its training with helium gave the crew greater assurance and extended their potential. Three hundred feet was no longer a redoubtable frontier. Falco's feelings are interesting in this regard:

"At 275 to 300 feet," he says, "with compressed air we were unable to do

Recco looks over the coral he has gathered during a decompression pause.

FENZY 70

the smallest job due to the narcosis that clouded our minds. Using helium kept our minds completely lucid."

In Corsica

Commander Cousteau wanted to make a film in the Mediterranean about coral and coral fishers. From July 2 to 10, 1971, Falco made a reconnaissance trip to Corsica, Sardinia, and Greece with Jacques Renoir. At last Corsica was chosen, particularly the Strait of Bonifacio. This choice was largely due to the dives made there by a champion—and incredibly bold—coral fisher named Tony Recco, a native of Propriano who stayed 300 feet down for 10, 15, or 20 minutes with bottles of compressed air and did not observe the decompression pauses very strictly. Others braved the same danger: another coral fisher, Gavinno, told the crew that returning from a dive to 340 feet he ascended 100 feet in three minutes. Six hours later his spine hurt and his feet were paralyzed. He had to spend time in a caisson. Afterwards he returned to coral fishing using the same dangerous techniques.

Corsicans have been coral fishing in the Strait of Bonifacio for a long time. They used to find coral at 120 feet, but they cleared that out and now have to find it between 250 and 300 feet. That is already very deep, and it is unlikely that any exists deeper.

Falco searched the area thoroughly in the diving saucer, and he never saw any good coral below 375 feet. It seems that beyond that depth red coral becomes tiny and then disappears.

Coral grows back when it is cut off, but biologists say it needs about 200 years to form a good branch. The Strait of Bonifacio is particularly rich because of the great depth. The coral there is very beautiful thanks to the very abundant plankton it feeds on. But the coral fishers who live so dangerously are taking the best clumps, and their descendents for several generations will no longer be able to admire these marvelous jewels of the sea.

Saint Andrew's cross

True, old-fashioned methods of coral fishing are even more murderous. Some fishers still use the Saint Andrew's cross that was already common in the fifth century B.C. That amounts to two planks spread with netting and used to scrape the bottom. Between Sardinia and Corsica there is a plateau in the sea where this archaic form of fishing is still practiced. The *Calypso* went there and Falco dived with the saucer to see the devastating cross in operation.

The plateau is about 250 feet in area and covered with sand. Splendid vaults have been hollowed out on the side of the plateau, and there grow large branches of coral. The finest are at about 375 feet and are protected from the cross by an overhang. But wherever it scrapes the bottom it creates havoc. Everything is crushed: coral, gorgonia, sea urchins. The coral that is brought up is only a fraction of what has been destroyed.

The tool now used is an iron cross dragged by a trawler manned by seven or eight fishermen. In the saucer Falco watched the cross tumble right and left, breaking the most beautiful clumps of coral. The little that stays caught in the net represents only 15 to 20 pounds, barely enough to earn a living for the crew.

With the diving bell

To film the coral fishers at great depth it was decided to use the Galeazzi bell at 120 feet. Reference tables for the helium-oxygen mix showed that the divers' stay should not exceed ten minutes.

The *Calypso* was moored to four big anchors in the Strait of Bonifacio, directly above the coral area reconnoitered by the saucer. When finished, the photographers entered the bell, closed the door, and phoned the *Calypso,* which hoisted the bell up and put it in the hold. They were then decompressed comfortably in the bell, where the breathing mixture was carefully regulated.

The problem was to be able to use the bell every day. When weather was good, it was easy. But when the sea was rough, the operation was tricky.

Bad weather could thus prevent them from filming Recco, who dived whatever the conditions. He used an inflatable boat with a big rock attached to a nylon cord as anchor. On the bottom he operated within a radius of about 100 feet of the rock. To find it easily he attached white streamers about 15 feet above the end that waved in the current. This reference point was vital for him.

He did not use helium because it is expensive and also because the decompression pauses are much more rigid. Besides, divers using helium feel the cold more; those on the *Calypso* had to wear several suits at once.

Nadine, Recco's "sailor," visits him during a decompression pause and breathes air from his mouthpiece.

During their decompression pause the divers breathe oxygen in the water: starting at 36 feet their regulators were fed oxygen from the surface.

At right: Back from a deep dive, the men enter the Galaezzi bell.

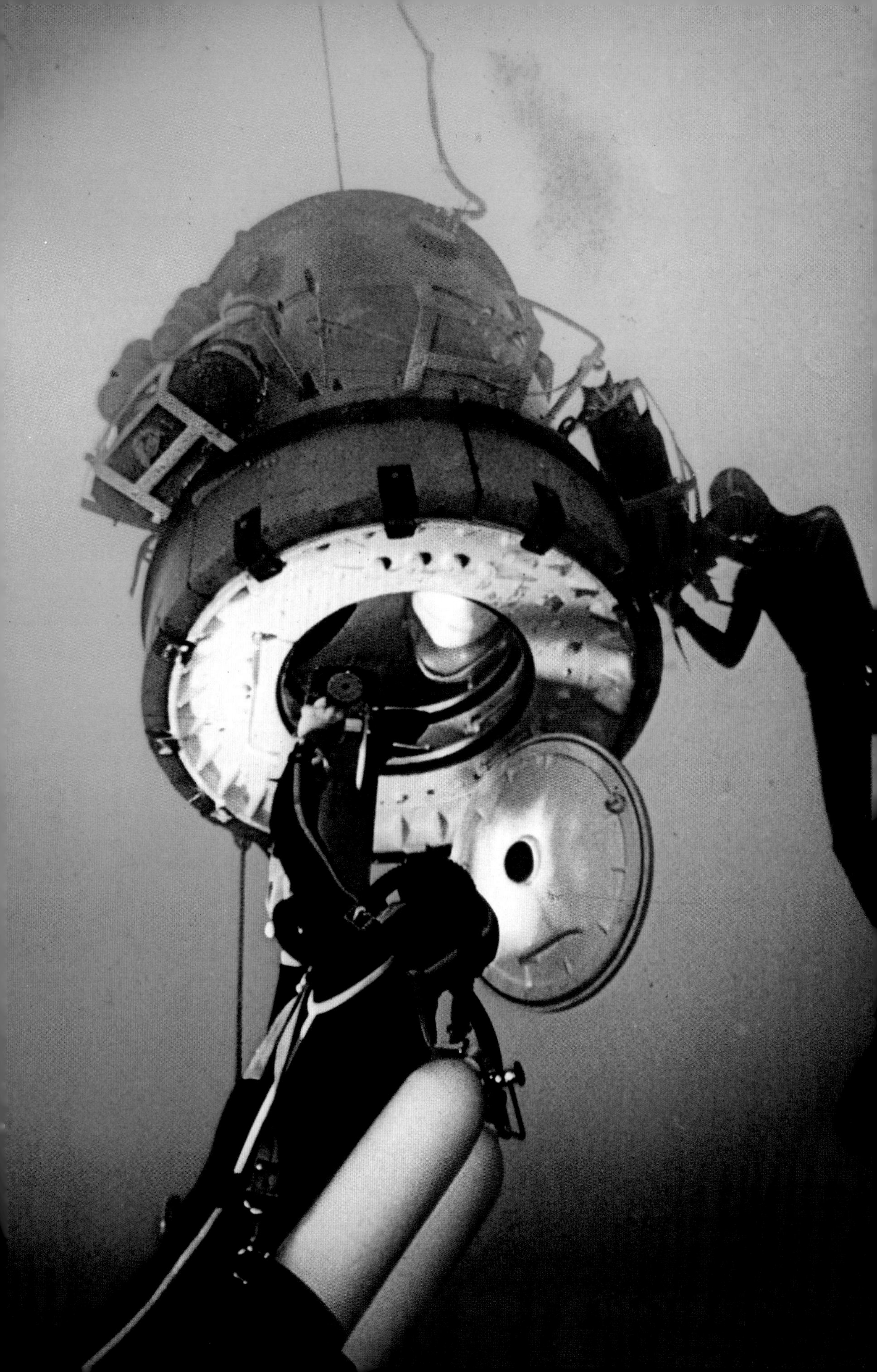

Recco could dive in any weather because he did not use a bell for decompression. So Falco decided that the crew would also do without one for its decompression pauses. But the problem was to breath oxygen under pressure during the pauses. That was easy in the Galeazzi bell, but in the open sea it was more hazardous. In his methodical way, Falco prepared a system that offered a maximum of safety: a graduated scale was submerged with three pauses marked on it beginning at 40 feet. Three oxygen tubes were attached to it.

''We rehearsed it,'' says Falco. ''Naturally I tried it first, and it worked well. It was important to remain as relaxed as possible and to move very little. But cold was our worst enemy. All went well the first three or four times. Then one day, maybe due to fatigue, Coll fainted at 40 feet. That day, unluckily, I had not dived. I was on the *Calypso*'s bridge. Fortunately the divers with him noticed he had fainted and brought him up. His teeth were clenched and there was foam at his mouth. I got his suit off and put him in the bell with the doctor. Ten minutes later, when Coll had just been put under pressure in the bell, I was alerted that a second diver, Christian Bonnici, had fainted: we had to decompress the caisson fast and put Christian in it too. Fortunately the pressure had only mounted to 300 grams, equivalent to 10 feet.

''The two of them were completely unconscious. Over the Galaezzi's telephone I could hear them moan and gasp. It was frightening, and I was so worried that I called Captain Alinat in Monaco, who reassured me that all we could do was put them in the caisson. They had probably fainted due to the oxygen—hyperoxygenation—and would come to slowly.

''He was right, and they were conscious 15 to 20 minutes later.

''Since they could not be brought to the surface the instant they fainted, they had absorbed some seawater. The *Calypso* sailed for Ajaccio, where they were X-rayed and treated. Two days later they were back to normal.''

Shooting the film

Every morning the divers, equipped with cameras and floodlamps, followed Recco as he descended along his nylon cord with a weight of about 60 pounds. To save time and effort he let the weight carry him down.

To film the descent it was necessary to leave the surface at the same time and go as fast as he did. Three divers hooked on 150-pound weights that were attached to the surface with a line. When Recco gave the signal, the line was cut and everyone glided to the bottom. One diver held the light, another the camera, and the third maneuvered the weight on the cable. The descent—there were over a dozen—took one or two minutes.

On the bottom the water was brighter—"like a dawn," as the divers said. The light came from daylight being reflected up and was tinted blue-green.

Recco appeared marvelously at ease at 300 feet down. He seemed completely unaffected by narcosis. He tapped at the base of the coral with his hammer, took the branch, and put it in a net hanging around his neck. Hammering at exactly the right spot, he lost not a second. This he did, despite being unable to use his legs, which were half-paralyzed as the result of a decompression accident. On land he limped heavily, and he swam only with his arms.

"If I'm ever unable to find my nylon line on the bottom, I won't be able to surface," he admitted. "My legs are just dead weight."

If the *Calypso*'s divers had breathed compressed air like Recco, they would not have have been able to shoot the film. But with their helium-oxygen mixture they could swim faster, pull the cables, carry lights.

Recco found his coral more easily with the floodlamps on, but he was afraid of being momentarily blinded when they were turned off. He could gather 15 to 20 pounds of coral in as many minutes. Sometimes he found branches weighing up to two pounds.

Recco surfaced slowly along his nylon line, using a Fenzi buoy. If he felt a bubble in a joint, he went back down and increased his pauses.

His system of pauses was ingenious. At about 75 feet from the surface there was a second nylon line to which a lead belt was attached. He hooked the belt to the lower sling of his air bottle, using it as a sort of ballast and keel. This helped him stay perfectly immobile in the frequent swells of the Strait of Bonifacio, attached to the nylon line. Then he gradually went up from one pause to the next. Since his three or four diving suits squeezed him badly, he did a virtual striptease, removing them one by one. Nadine, his assistant, picked them up on the surface.

In all, his pauses lasted between one hour 40 minutes and two hours, like those of the *Calypso*'s divers. But he did them in the water, whereas since the accident with Bonnici and Coll, the bell had been put back in service to insure safe decompression.

Blossoming polyps of red coral in the Mediterranean.

Returning from a deep dive, the men remove their bottles and enter the Galeazzi bell for decompression.

Recco's "assistant," his sailor as he called her, was a beautiful girl who watched over his dives from a boat. She always knew where he was and what he was doing. She watched him work by using a simple bucket with a glass bottom. She was convinced that one day he would not make it back. Each year some coral fisher dies in an accident in the Strait of Bonifacio. This girl dived often and went to visit Recco when he made his pauses. He passed her his mouthpiece and they took turns drawing on the air bottle. She took his net full of coral and his tools. Often he signaled to her that he wanted to see the best branch he had got that day. She found it and showed it to him. What did he think of as he looked at it? Of the money it represented? That he had risked his health and life? He loved coral passionately. On land he spent hours contemplating all the magnificent branches he possessed.

When the *Calypso*'s crew made his acquaintance, Recco made only one dive per day, but previously he had made two, which meant he had accumulated nearly a ton and a half of coral, which he piled in a shed. The year before he had sold enough to make $40,000. He lived very simply in a small cabin beside the water. His greatest joy was to open the door of his shed and say, "Look at my coral."

Recco took other risks besides not respecting decompression pauses. He also put 550 pounds of pressure in his air bottles to have more when he dived. He thus endangered everyone standing near his compressor. In fact he did have an accident: a bottle exploded, but luckily no one was hurt.

Falco took Recco with him for a dive in the saucer.

"We found some great coraling areas," Falco says. "Generally coral is on vertical walls or on an overhang or in the hollow of a rock. I also have seen coral planted like little trees, but very rarely. There were not many fish, only a few big wrasses weighing about two pounds, and rockfish. On the way down we saw a school of silvery pompano circling the divers. On the rocks were deep sea urchins, Echinus melo, the size of small melons. We saw the coral from afar, thanks to the open polyps that covered them like white flowers. When we came closer the polyps closed and uncovered the red coral branch.

"Recco was all eyes. He had never seen coral illuminated. He said, 'Let's go back up; I've got to dive down here.' He could size up the area fast. After five minutes of looking he said, 'There are 100 pounds of coral here.'

"Such reconnoitering helped set up each sequence of the film. I showed Recco the narrow fault line where he was to swim by during shooting. At another spot he was supposed to hold out his arm, use his hammer—and above all not look at the camera. He was always ready to do exactly as we asked."

Such is the story of the coral fisher of the impossible who disdained the laws of physiology and ignored fear. Today a film bears witness to his crazy audacity. But he is dead. Not from a diving accident, but by the hand of his brother-in-law during an obscure family quarrel. Spared by the sea, he was killed by a rifle shot. Poor Recco.

The *Calypso* in the straits of Gerlache in Antarctica. Note the helicopter pad on the bow.

10

beneath the ice cap

FIRE ABOARD—FIRST ICEBERG—CHRISTMAS DIVE —ON DECEPTION ISLAND—THE CRUISE CONTINUES —ON THE ICE CAP—KRILL—THE *ICE BIRD*—BLIZZARD—CALM—DAMAGE —PRISONERS OF THE ICE—SATELLITE WEATHER REPORT —DRAKE PASSAGE—AT 6 KNOTS IN HEAVY SWELLS

On November 1, 1972, Falco leaves Orly airport for Rio de Janeiro and Buenos Aires, where he joins the *Calypso.* As always, she looks small surrounded by larger ships.

There are still tons of matériel to load before sailing for the Antarctic. He helps cram it on the foredeck, which is already crowded with the helicopter pad over the bow. Five big crates from NASA sit on the rear hold hatch, receivers for satellite photos of areas where the *Calypso* will be working near the South Pole.

The Calypso sails November 6 with two American specialists from NASA aboard.

The ship rolls heavily and it is cold. The sky is constantly overcast. Falco and his comrades try out their down bedcovers and pass an excellent night. On following days the crew attempts to approach some whales sighted by the helicopter off the point before the Valdés Peninsula, and the film crew shoots some good underwater sequences.

On November 28 the ship passes the black, rugged peaks of Staten Island where there are patches of snow. It heads for Ushuaia, where it takes on bottles of butane for inflating the hot-air balloon.

Fire aboard

"On October 5 we exit from the Beagle Channel," Falco recounts. "At about 4 P.M., while I am on the bow, I hear Madame Cousteau shout 'Fire!' I run for the rear and I see some crates flaming on the port side. I throw one in the water. Morgan, the cook, gives me a pan of dishwater that I throw on another crate, and Jean-Marie France, the mechanic, helps me to push it overboard. It's all over by the time the extinguishers arrive.

"Were were lucky: Some gasoline was nearby, but the only damage was to the paint on the gunwale and the bridge—souvenirs of an incident that could have been the end of the *Calypso*.

"We take up a southerly heading. At 3 A.M. I am awakened by heavy rolling. I get up. There is no wind, but two swells, one from the southwest and the other from the northwest, are hitting us. The air temperature is 37 degrees, the water temperature 35, and the humidity is 90 percent. We begin iceberg watch. At 6 A.M. Dominique Sumian's down jacket and face are covered with frost from thick fog during his watch. The crane, guy-wires, and halyards are all covered with ice.

"The fog is gone by 8 A.M. and a wind comes up from the southeast. Toward 4 P.M. it reaches force 8. Due to the storm, the Commander abandons his idea of going to Elephant Island and heads for King George Island. By late evening the sea is very heavy and the port door on the bridge is broken in two by a wave."

First iceberg

"At 2:30 A.M. on November 8 I take my watch with Commander Brenot, and I notice the first blocks of ice on the sea. I see several icebergs ahead of us on the radar scope. When Bonnici relieves me, I see straight ahead the first iceberg, huge and impressive. It is sometimes white, sometimes blue or green. The iceberg is drifting against the wind, throwing up great waves. It is snowing.

"At 7:30 we anchor in the shelter of King George Island. It is dead calm. I sight the whale fishers' sheds, two whale boneyards, and four tombstones. Five hundred yards to the north a large glacier is about to slide into the sea. It will be just as well for us not to be on hand when it gets going. Several penguins and a seal look us over from the shore.

"Toward noon, while we are in the mess, a block of ice drifts alongside the *Calypso*. It is occupied by penguins that seem not the least concerned by our presence."

November 9. "It has snowed all night. Blocks of ice come and go in the bay, depending on the wind. I decide to dive to try out an inflatable suit. It is too small and it's hard to get my head through. I swim beneath the *Calypso* at 100 feet down, where there is only black mud and five feet of visibility. I surface with a little water in the left glove, enough to freeze my hand.

"That afternoon we all go ashore and reconstruct a 75-foot blue whale with bones found there. Dr. Raymond Duguy shows us where the bones go.

"At about 6:30 P.M. we go back aboard and learn that part of the helicopter pad has fallen into the water. Yvan Giacoletto dives but is unable to find it. His regulator freezes at 115 feet, and he comes back with a bleeding ear."

November 10. "I dive to look for the piece of platform, but in vain. Between 100 and 135 feet I can make out red tubular sponges, greenish urchins, and brittle stars. I bring back a sponge with the rock it was on. When I put it in the launch, a small fish with green on the top of its head slides out. Is the Antarctic the land of colorful animals?

"At midnight there is still daylight."

November 12. "At 7 in the morning the wind is blowing at 50 knots. The *Calypso* is dragging its anchor. The helicopter has to be tied down. The anchor with four links gets caught as we raise it, and we have to use the crane and winch to get the rest up. We are all soaked to the skin, and the water hitting us in the face is glacial.

"The Commander keeps the *Calypso* stern to the wind."

November 13. "Weather changes fast here. Brilliant sun today. Westerly wind, hardly 20 knots. We head for the west shore of the bay. Three teams go

Double page following: A diver looks over an iceberg.

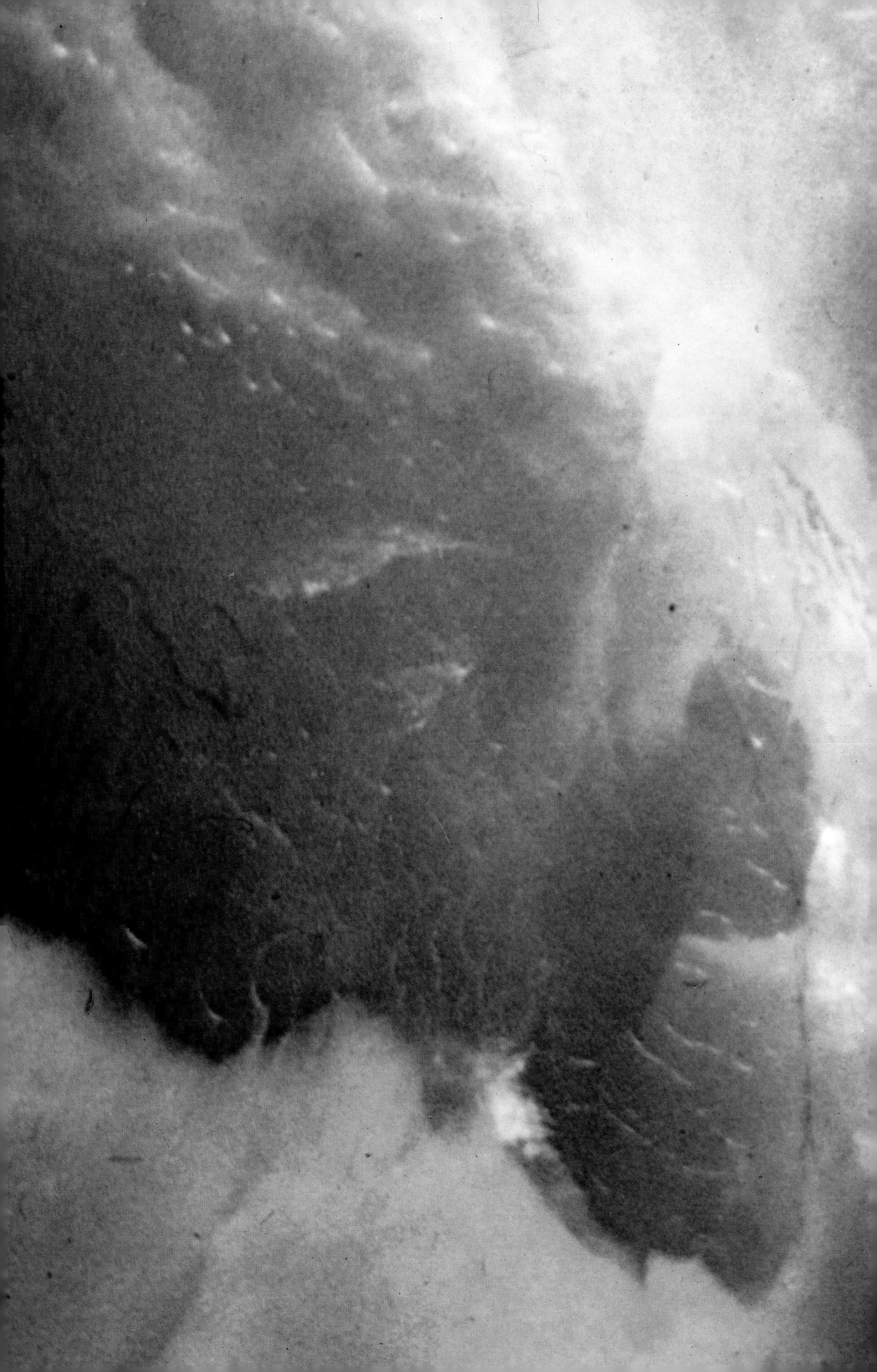

ashore. Mine is assigned to film giant petrels. I pet one for 15 minutes after giving it bits of ham.

"At 8 P.M. the weather is good and the Commander decides to weigh anchor.

"We leave behind immense icebergs riddled with grottos. Low clouds hide all but the bases of D'Urville and Joinville.

"At 6:30 next morning we come to an ice barrier and can see smoke from the cargo ship *Bahia Aguire,* anchored three miles away at the Petrel base on Dundee Island. We moor next to her after weaving between blocks of ice. The Chilean cargo ship has brought us the diving saucer and Gaston, the mechanic in charge of its maintenance. We take on 40 barrels of airplane fuel, diesel fuel, and some food.

"At 6:30 P.M. we slip between blocks of ice toward the open sea. It is dead calm.

"I get up at 10:30, unable to sleep. It is broad daylight. I go to the bridge and see some icebergs whose transparent bottoms are shot through with holes. We come across a seal sleeping on an ice floe; its body heat has hollowed out the ice in its shape.

"According to Captain Bougaran's figures, an iceberg 600 feet on the side like the one we just encountered would weigh 8 million tons.

"We drop anchor in Peters Bay and spend several days observing and filming two species of penguin.

"Numerous young are in nests made of pebbles. They clamor for food. Their parents regurgitate a meal of krill, the whale food that is very abundant in this region.

"They fear the large sea eagles that dive suddenly on the nests. At the edge of the beach a leopard seal is ready to ambush a penguin returning tired from food gathering."

Christmas dive

On December 20 the *Calypso* reaches Deception Island and then the volcanic island of Chaco, where it shelters from the wind. With its hippopotamuslike silhouette, this island looks terribly hostile and aggressive: it is surrounded by a black belt of rocky points.

"We have champagne with our Christmas dinner aboard the *Calypso.* It's summer here, and yet it's snowing."

Falco, who has seen so many seascapes in the diving saucer, including the Arctic Ocean off Alaska, now visits the underwater world of Antarctica. He dives in the saucer on December 25 with François Charlet as passenger, and they touch bottom at 250 feet. The water is clear, whereas it had been cloudy with plankton on the surface. Visibility is 25 to 30 feet. There is a volcanic mud bottom, with an extraordinary abundance of brittle stars, which seem to feed on krill. There are a lot of little fish resembling gobies. A big anemone is eating a smaller one. Large bouquets of transparent ascidiums rise from the bottom with small sea urchins attached to them. Large quantities of shrimp are attracted by the saucer's floodlamps. Marine life appears very rich in this part of the Antarctic.

At 6 A.M. on December 26 the *Calypso* weighs anchor to sail for an underwater volcano southwest of Deception Island. The saucer is launched. At 210 feet the water is clear and on the mud slope Falco notes thousands of transparent ascidiums. He continues east to the foot of the volcano. Along the way he encounters large brown and white flat worms enlaced, a few orange and yellow sponges and more ascidiums down to 300 feet. There the brittle stars are predominant.

On Deception Island

Tragedy strikes on December 28, 1972. Falco's notes read:

"At 6 A.M. I join Michel Laval on the bridge. He talks nostalgically about Marseilles, about how much he hates these cold regions. He is first to take off in the helicopter. I follow it with binoculars. It lands on the glacier, comes back to pick up Colin Mounier and Yvan Giacoletto, and takes off again.

"The helicopter pilot calls on the radio. He sounds panic-stricken: Michel Laval has been struck by the rear propeller. Fifteen minutes later we unload him from the helicopter on the front platform. I help carry the stretcher. His skull has been bashed in.

" 'It's all over,' the doctor tells me. 'Probably just as well.'

"The Commander decides to take Michel Laval back to France himself. The *Calypso* takes him to Ushuaia. A crew of eleven will stay here in what is left of

the English base, which has been abandoned since the last eruption of the volcano in 1970. I am one of those who stay.

"We spend the day unloading two and a half tons of matériel for the stay. That night I sleep aboard the *Calypso*. It is past midnight when I fall asleep, exhausted.

"Next day we unload four barrels of 50 gallons of gasoline. At 8 in the morning the *Calypso* enters Deception Strait, then disappears.

"We set up our base. Prezelin clears out around the building with a shovel while I put a new smokestack on the old cookstove, which is soon roaring. We have plenty of coal, dozens of bags left by our predecessors. Jouas gets his radio working and at 4:45 P.M. picks up the *Calypso*.

"Outside the wind is blowing 30 knots. The sea is gray, the clouds low. Penguins are sleeping on the shore. I go fishing and bring back a good catch.

"Our stove goes out during the night. The cabin is terrifically cold and everyone feels it. That will teach us to watch the fire all night.

"We've discovered that in Pendulum Bay the water has been heated by a volcano and is lukewarm. We go to have a swim, along with the filmmakers. But we get a bad surprise: only the top few inches is warm, all the rest is icy and we come out frozen. I make a big fire with driftwood.

Opposite page: Launching the S.P. 350 saucer near an iceberg.

Three views of the Antarctic bottom taken from the saucer: the animal with long legs is probably a Pycnogonide; a nearly transparent squid, and an ascidium on a bottom covered with urchins.

"Our New Year's Eve menu is penguin with peas. It's tough and smells of fish. We had picked it up just after it had been slashed open by a leopard seal.

"On New Year's Day we are hit by a raging blizzard driving loads of snow.

"On January 3 the weather is good and the wind dead calm.

"At 9:30 that morning Philippe Cousteau, Dominique Sumian, and Christian Bonnici dive near an iceberg and return enthusiastic: the sun on submerged ice is a fantastic sight.

"At 4 P.M. I dive with a scuba and run along a cliff face at 60 feet covered with sponges, algae, ascidiums, and anemones, along with many fish. I find the wreck of a whaleboat, the *Southern Hunter.* I look it over and find an explosive harpoon head on the bridge that I bring back.

"At 9:30 P.M. on January 4 we set off in two inflatable boats. In front of the pass we sight a large flat iceberg far off. I estimate it to be eight miles away. We make for it, and after a half hour I realize that it is much farther away than I thought. Light in the Antarctic is tricky. Fortunately the water is calm and we are warmed by the sun. It takes one hour and ten minutes to reach the iceberg, an enormous mass 300 yards long, 100 wide and 40 high. Through the blue water we can see the submerged part going down over 180 feet. There are little grottos on the shadow side with thin stalactites hanging down. Everyone in the boats is quiet, absorbed by this extraordinary spectacle, moved by the immobile beauty of the sea."

The cruise continues

The *Calypso* returns to Deception Island on January 7 and embarks its crew members, who are in good health and spirits.

The *Calypso* goes first to Melchior Island and the American base, Palmer, through the Straits of Gerlache, then makes for the south amid an increasing number of icebergs.

Falco is awakened at 3 A.M. on January 17 by the sound of ice banging on the bow. Through the porthole he sees the ice pack that surrounds the *Calypso.* He wakes Commander Cousteau, while Commander Brenot on the bridge turns the ship around. The way is blocked; the ice floes and bergs make every maneuver dangerous. Two hundred yards ahead, Falco sees a reef and an island.

"There's an error in the charts," Cousteau says.

He circles until 4 P.M., then decides to head north, even if it means breaking the ice with the ship's bow. At Hommack Island there is a wall of ice. The helicopter goes up to reconnoiter, and the pilot returns with the news, "The only way out is to the south." Several floes strike the *Calypso* and its screws.

Toward 9 P.M. a big iceberg moved by the current is detached from the pack and opens the way, letting the ship out of what was very nearly a trap. The wind drops and the sea is calm.

During the next two days of good weather the ship goes along the pack and reaches Rabot Island.

Several divers explore an immense iceberg, swimming around and within it. Grottos have been cut out of the ice, making pockets of water where the divers take their lights, which create blue reflections in the ice and gleams in the stalactites. It is an unforgettable experience, but not without its dangers: at any moment unstable blocks of ice could fall and imprison the divers in the pockets, or the whole iceberg could collapse on them.

During Falco's watch on January 20, he counts over 150 icebergs around the ship. The next day the *Calypso* penetrates into the ice pack and is moored securely to it. The weather is splendid, and the Antarctic looks radiant. The hot-air balloon is filled and Philippe Cousteau takes it up with Louis Prezelin.

The same day the saucer dives alongside an iceberg in Margaret Bay. Falco descends to 250 feet, where he finds an ice overhang carved with large ridges. Little transparent fish live there—"ice fish" whose blood contains a sort of antifreeze. The bottom at 700 feet is rich with life: gray sea cucumbers, large tubular sponges, Bryozoa, Annelida, strange little balls supported in the water by orange strings. There are also an enormous sea star with over 20 arms and a school of silvery fish. The saucer surfaces at midnight amid icebergs whose bases leave deep furrows in the bottom.

On the ice cap

After a reconnaissance flight in the helicopter, Cousteau decides to approach the ice cap through narrow channels bordered by glaciers. The *Calypso* sails along Hansen Island, whose 3,000-foot peaks are covered with ice. The channel is barely 750 feet wide and several times white cliffs plunge into the sea with the sound of thunder. Crew members in the launches use gaffs or their hands to push floes out of the *Calypso*'s path. The tiring six-hour run ends beside the ice cap, where the *Calypso* moors.

Here is Falco's diary of that busy day:

"Dozens of seals are lying on the edge of the ice cap. At 2 P.M. a team leaves with two sleds for the seal holes about 650 yards to the north. I go with Raymond Coll to film the seals in the water, but when they see him and François Dorado under the water they shoot toward surface. Too bad: the water is as limpid and blue as the Mediterranean.

"Our return to the *Calypso* is complicated by the ice floes in our way. These bits of the ice cap have floated south in just two hours, and I realize that the exit channel is slowly closing. I notify the Commander that we should weigh anchor, but the last sled doesn't get back until 6 P.M. The saucer dives once more, breaking through the gradually forming ice, and then we sail.

"At 7 P.M. the two launches open the channel ahead of the *Calypso.* This struggle to get out lasts until 1 A.M. and I go to bed exhausted. The night before I had only one hour's sleep. My face is sunburned and my hands bloody. The sky is overcast and snow is not far away."

January 25. "I wake up this morning at 10 o'clock after taking last night's 1 to 4 watch. It's overcast and there's a 20-knot wind. In the helicopter, the Commander has found ice everywhere to the north of Adelaide Island, and the *Calypso* can't go back toward the ice pack. The Commander decides to go northwest of Adelaide to make some dives around the icebergs. With the *Calypso* anchored 500 yards from the glaciers above a 60-foot bottom at 2 P.M., we get ready. Snow starts falling and the wind from the glaciers freezes us as we harness up on the after deck.

"The icebergs are blocked by the shallow bottom, which makes me feel better. The saucer will run less risk of a collision with them.

"The launches pull a few ice blocks out of the way and then the saucer dives. At 60 feet I find the bottom, then move deeper. At 120 feet there are plaques of sponges, at 275 feet ear-shaped sponges, sea stars, gorgonia. At 300 feet there are numerous brittle stars, some quite large, and ice fish. While surfacing in the saucer I hear the awful noise of a rending iceberg."

Krill

The *Calypso* starts toward the south to go around the ice pack and then heads

Two men in an inflatable boat approach a shore where penguins are taking a sun bath.

On Adelaide Island Albert Falco photographs a seal that Commander Cousteau is trying to approach. François Charlet is handling the camera and Guy Jouas the sound.

north. The sky darkens and the sea becomes heavy. Albatross cry in the wind.

From the helicopter pad Falco sees thousands of little crustaceans in the water: krill, the favorite food of whales. Commander Cousteau decides to send the film crew down, and for an hour they maneuver through a school of shrimp. Falco gets several pounds of them with a hand net and the cook fries them for dinner.

The *Calypso* makes it to Palmer base, but it has a broken screw. The divers spend two hours beneath the hull attempting to take it off, and finally have to tie it firmly in place. The weather worsens to a storm during these two days, but the *Calypso* is sheltered. An unexpected visitor shows up.

The Ice Bird

January 29, 1973. "I wake up at 4 A.M. Something knocking on the *Calypso*'s hull makes me jump out of my bunk. I see a blue sailboat with a strange mast anchored between us and the rock. A man with a beard who appears exhausted and soaked from head to foot is paying out an anchor on the bow. But the anchor slips and the boat is about to run into the rock. I take the launch and pull the sailboat over to moor it with the *Calypso.* Then I give the man some coffee, which he drinks avidly. Water is running from his clothing and his fingers are so beaten up that his nails will probably fall out.

"He is a single-handed sailor, David Lewis. His mast broke on November 29, 1972, and since leaving New Zealand he has been sailing toward Antarctica with a makeshift mast fashioned from a boom. His boat turned over. He managed to right it again, but everything that stayed aboard has been damaged by water.

"We make some photos of him and his boat, the *Ice Bird,* which will illustrate the book he hopes to write about his voyage.

"We leave Palmer at 3 A.M. on January 30 and head for Paradise Island. Along the way we test the left screw, which has been straightened out, running it up to 840 rpm with no vibration. From 840 to 940 rpm there is some light vibration. We make the trip in four hours at ten knots."

February 1, 1973. "At 7 A.M. we are close to Fleires Island in the Straits of Gerlache. The *Calypso* moors to the island with a nylon line attached to a stake.

"Toward 10:45 the Commander and I go down with the saucer along a cliff to shoot some film. We count five or six different species of shrimp, including krill. Animal life is quite varied: giant sponges, white coral, crabs with very slender legs. We see more large sea stars with many arms that absorb the shrimp. A strangely shaped long orange worm also catches shrimp by wrapping itself around them. When the shrimp is caught, the knot moves it to the creature's mouth.

"A pear-shaped ascidium also traps shrimp with a double horn with retractible filaments. A pink fish with a very sharp tail looks like a nail.

"While surfacing we encounter numerous pink urchins, and at 60 feet a forest of large seaweed.

"Today we certainly have seen a strange world."

February 3. "Heavy swells wake me at 8:30 A.M. as we are southwest of Melchior Island. Guided by the helicopter, we make for the Dufaur Islands where, after skirting some dangerous reefs, we drop anchor near an island where the Commander has spotted some sea lions, which are supposed to be non-existent in Antarctica. A crew takes off at 2:30 to film them. I go with them and land on the island to try to chase about 20 sea lions into the water. They bare their teeth at me, but finally consent to go toward the water. Philippe Cousteau and Raymond Coll film that, and then we all get down on our knees to try to touch a sea lion, but none of us gets closer than ten feet."

Blizzard

On February 9 the *Calypso* is in Esperanza Bay amid numerous chunks of the ice cap mixed with large icebergs. But all is calm. Some young penguins, watched by a few seals, slide around or fight to see who will stay atop a hillock of ice.

In the afternoon snow falls abundantly and covers the *Calypso* with a thick layer. At 7 P.M. a glacial wind from the south comes up and the floes start toward the ship.

Paul Zuena pushes the first ones away with the gaff. But there are more and bigger ones coming. The wind is at 25 knots. The blizzard blinds those trying to protect the *Calypso.* A big block of ice is caught on the bow. Groups of penguins squeeze against each other for warmth. The 40-knot wind raises clouds of snow. The motors are started and the anchor starts up. Just at that moment a large ice floe hits the *Calypso* on the port side and makes a hole in the outer hull.* The anchor is finally raised and the *Calypso* makes it to the shelter of a glacier at the end of the bay. The snow is getting thicker on the bridge.

Falco goes to the aft hold to repair the damage done by the ice floe. The mechanics inform him that a cylinder head spring on the port engine is broken.

''I finally get to bed at 2:30,'' Falco writes, ''with the wind still howling. I think it will be a short night.''

February 10. ''I was right.I am shortly awakend by the radio operator, Paoletti. The Commander is doubling the watches, and I have to go to the bridge. I dress quickly and step out into 15 inches of snow on the starboard gunwale. I find the Commander and Captain Bougaran on the bridge. They are keeping the *Calypso* 150 feet from the glacier by reversing first the starboard and then the port engines. The Commander says, 'I call on the old guard when things are bad.'

''Bougaran stays until 3 A.M. with me. At the end of his watch the wind is at 40-50 knots with gusts to 60. Yesterday's snow has hardened, and the ship is covered with ice.

''Several large floes pass aft, and one hits us, fortunately, without any damage.

''At 5 A.M. I wake the Commander. The wind is still strong. Dawn broke at 3:30 and I can keep a watch on the stern through the doors of the bridge. Sometimes we get within 75 feet of the rock.

''At 10:30 the blizzard and sea spray have accumulated tons of ice on the afterdeck, covering the saucer, the launch, the crane, and the inflatable boats. I resume my watch at noon. We alternate every two hours with Commander Brenot and keep the *Calypso* in reverse against the wind.''

*The *Calypso,* a former mine sweeper, has a double hull of wood.

The launch clears out ice from the *Calypso's* path.

Albert Falco on the *Calypso*'s ice-covered afterdeck.

Calm

The next day at 7:30 A.M. the sky is covered but the wind has dropped. The ice floes are herded together at the entrance to the bay and seem to wait for a signal to charge again.

Aboard the *Calypso,* where the crew has the impression of being under attack by an enemy army, work starts again with filming on the afterdeck. Falco takes an inflatable boat and shoots the imprisoned *Calypso* with his 8 mm camera.

Along the way he glances at the port hull aft and notices that the hole made by that block of ice is deep. After lunch, Guy Jouas and Falco film some young penguins that they push into the water, diving after them into a tunnel in the ice. But it starts snowing hard under a south-south-west wind and everyone goes back aboard to warm their frozen hands.

Damage

Falco resumes his account:

"I am awakened at midnight on February 12 by the noise of the anchor. I join Commander Cousteau and Philippe on the bridge, and Bougaran is in the map room checking the satellite weather reports. The wind is blowing 40 or 50 knots. It's still snowing. The *Calypso* is once again cruising back and forth in Hope Bay. At 3:30 A.M. Commander Brenot tells me that the port engine is not powering the screw.

"At first I don't understand, but then I realize the problem must be with the propeller shaft, following our collision with a block of ice. The chief mechanic confirms my suspicion: the port shaft is out of whack. The Commander is worried and decides to drop anchor in 20 fathoms. I offer to go look at the damage because he hesitates to ask me.

"In the water I have to fight to keep from being thrown against the hull, but I can see immediately that the shaft is broken in the sleeve. I also discover that two blades of the other screw are twisted at the ends.

"I take advantage of this unexpected dive to descend to the bottom at 90 feet. There is a rounded rock covered with a thin layer of mud and brittle stars. It's not surprising that our anchor does not hold on this bottom. But my regulator freezes, partly cutting my air flow, so I go back up hoisted aboard by the crane as the sea lashes me.

"The crane deposits me on the afterdeck amid a group around the Commander. My news is all bad: broken shaft, propellor blades twisted. . . ."

Prisoners of the ice

The *Calypso* has to wait for better weather to venture out of the bay in direction of the nearest base, Petrel.

At 3 P.M. a diving team puts the propeller in a sling so it will not get in the way of the rudder. Dominique Sumian comes up with his sinuses blocked and his regulator full of ice. On deck the thermometer reads twenty-five degrees and the wind is 40 knots. Philippe, who has filmed the underwater work, also has his sinuses blocked. Bonnici and Coll finish the sling and are the last back.

The crane breaks down when we try to put the saucer in the hold, and we postpone that operation. The Commander decides to send the helicopter to Petrel to tell them we are coming for repairs. In any case the *Calypso* has to spend the night in this trap. At the northern end of the bay large icebergs are ready to assault the ship if the wind turns.

The wind drops on February 13 and we can get the saucer down into the hold. The launch and the two inflatable boats take the divers to the northern point of the bay, but ice surrounds the three boats and they barely get away.

The dive is made at the foot of an iceberg of blue ice hollowed out into beautiful grottos with translucent, mirrorlike walls. When the divers surface, an ice overhang falls a few feet from Philippe Cousteau and François Dorado.

At 3:30 P.M. the Commander asks Falco to go explore possible passages through Hope Bay in the helicopter. He finds that the southern end is covered with a continuous field of ice toward Petrel. To the north, behind Bransfield Island, there are ice floes to the horizon. To the west he can see some blue water passages between icebergs. The sight is impressive and discouraging. The masses of icebergs in the shape of pyramids or sugar loafs stand like challenges to the men of the *Calypso.* At their base the ice takes on a marvelous sapphire-blue color or emerald green.

Back at the ship, Falco tells Cousteau what he has seen, and the Commander

The *Calypso* sailing west of Adelaide Island.

Above right: The play of light on an iceberg.

Below right: Men of Cousteau's crew contemplate a majestic landscape from the helicopter pad.

decides to leave that evening for King George Island. The danger of being trapped has never been so great.

Satellite weather reports

"We weigh anchor at 6:30 P.M.," Falco says, "among the icebergs. I take the watch until 1 A.M. and weave along trying to keep a heading of 320. Fortunately it is now dead calm, and the moonlight helps me see any floes that could be fatal to our starboard propeller.

"At 1 A.M. we learn from the satellite report that there is a low pressure area to the north of Drake Passage and that there will be strong northeast winds tomorrow. Luckily we are out of Hope Bay. We anchor at the Russian base to the west of King George Island.

"The worst is still ahead: Drake Passage, where the *Calypso* with a single damaged propeller could be surprised by bad weather."

February 14. "It's snowing and the northeast wind hits us. If our only propeller and only motor fail us, we'll have to hope to be tugged by a ship in the area. The Commander decides to dismantle the helicopter pad over the bow. Paul Zuena rigs up the nylon tug line that moored us to the laboratory buoy. The two forward anchors are taken out and a double chain passed through the hawse-hole. The nylon line will be attached to that.

"The wind screams again after barely 24 hours of calm. The Commander decides, at Bourgaran's request, to wait for the latest satellite information before sailing.

February 15. "During my 3 to 5 watch, Jouas tells me that the U.S. Navy and the Coast Guard are aware of our situation and are ready to escort us.

"We spend the day at the Russian base while waiting for the weather report. I return to the ship at 6:30 P.M. and Brenot tells me that we will be able to make Drake Passage tomorrow. We will have to sail at 5 A.M. I hope we can, for we would have to wait until February 19 for another chance.

"At 8 P.M., while 14 men are still on shore, a strong wind comes up. The anemometer goes from 10 to 45 knots. The crew ashore tries to get back to the ship in the inflatable boat, but is thrown back twice. Philippe, Bonnici, Sumian are soaked. The Russians give them an alcohol rub to warm them up, then bring them back to the ship in an amphibious craft just as Coll, Dorado, and I were getting ready to go for them.''

Drake passage

February 16. ''The Commander asks me to come and discuss making a floating anchor with the help of J.M. France, chief mechanic. We can make it with pieces of the helicopter pad if we don't sail today. Given the *Calypso*'s condition and the danger of its only motor stopping, a floating anchor would enable us to survive a storm. It's an old trick we have already used in the Mediterranean.

''But at noon Captain Bougaran returns from shore with a good weather report. Of the three low pressure zones threatening Drake Passage, one is heading east and another nearer us is dissipating. It's agreed that we will sail at 2 P.M.

''Instead of making the floating anchor from the helicopter pad we decide it will be faster to use the accommodation ladder and 300 feet of netting. Three hours later the anchor is ready.

''A whale passes on the port side. The southwest wind falls off and the sun warms us. It feels as if we are leaving the icebergs and glaciers, heading home. I'm sorry not to have dived more often, but the cold limits that. I can feel this land's hostility to man in my bones.

''At 8 P.M., just as I am having a last look at Nelson Island, I see two helicopters coming at us from the west. We soon recognize them as being from the Argentinian icebreaker *Piloto Bardo.* They hover off the *Calypso*'s port and starboard sides and and ask us by radio to contact Commodore Le Mée on the *Piloto Bardo.* A pleasant surprise: the Commodore informs the Commander that the *Yelcho* will rendezvous with us tomorrow to escort us. Everybody on board suddenly feels better.

''The helicopters leave at 8:30, as an enormous swell from the north makes the *Calypso* roll heavily. A few seals show up, then dive for the 550-foot bottom.

''Shortly afterwards we are surprised by swells from the southwest that make the *Calypso* creak and vibrate all over. She too seems tired. The gunwales are full

A diving team leaves in bad weather.

Above right: Divers attach the propeller shaft so it will not block the rudder.

Below right: Escorted by the Argentinian warship *Yelcho,* the *Calypso* sails through Drake Passage.

of water and it rains in our cabin, where it's impossible to sleep. I stay on the bridge.''

At six knots in heavy swells

February 18. ''The press has published accounts of the *Calypso* 's troubles, but the Commander refuses radio interviews for the moment.

''At 8 A.M. we are making six knots and it is raining. On our heading of 331 we have a headwind of 15 knots.

''Whereas we had expected the *Yelcho* at 3 P.M., it comes out of the fog behind us at 9:30 A.M. and slows down to escort us. None too soon, for the port engine is making noises that worry chief mechanic Jean-Marie France.

''At noon we are in the middle of Drake Passage. The American weather bureau forecasts a westerly wind of 20 to 25 knots for tomorrow.''

February 19. ''Six A.M., heavy swells from the northwest make the *Calypso* roll and creak as the wind goes from 18 to 25 knots.

''At 7 P.M. two Argentinian tugs take up positions near us, ready to intervene if necessary.

''At 8 P.M. the sea is even rougher, with dark clouds on the horizon. We have refused the tugs' offer of help.''

February 20. ''A powerful swell wakes me at 4 A.M. I look out and see Cape Horn under a lowering sky and, nearer, Deceit Island. A slight northerly wind carries with it the smell of the islands. Everyone on board grins and horses around. As for me, I feel as if I'm landing on another planet.

''At 9:30 A.M. the *Yelcho* signals the last Argentinian tug that it is in Chilean territorial waters, and the tug leaves. We would have been just as happy if it had stayed. What if we have a breakdown in the Beagle Channel?

''We enter the Murray Canal at 4 P.M. with two fast patrol boats escorting us. On the shore, a man waves a Chilean flag to greet us.

"We enter Beagle Channel at 8 P.M. and meet up with the two Argentinian tugs, which take up where the *Yelcho* leaves off. They escort us to Ushuaia, where we arrive at 11:30 that evening. It seems that Chili and Argentina are fighting over the honor of saving us."

11

outlook for the 21st century

FAMILIAR ANIMALS—THE TRIGGERFISH—DOLPHINS AND LIBERTY—APPROACHING WHALES—CHASES—AMONG FRIENDS—EXPLODING BOILERS—BRIEF VISITS—THE *RÉGALEC*

In the final analysis, Falco's whole life has been an attempt to approach marine life, hidden from man for centuries and revealed at last thanks to the scuba, the diving saucer, and the bathysphere.

All his physical exploits, all the risks he took had the same objective: to get to know the living beings that heretofore had been beyond the reach of man. Whether being jostled by a rough sea, diving in the darkness of great depths, staying in an underwater house or racing across the sea in an inflatable boat, Falco's motive is to touch a whale, have a shark eat out of his hand, pet a dolphin, tame a grouper, be followed by a triggerfish or even to learn the secrets of the life of a coral reef.

It would be wrong and unfair to think that this exploration of the depths de-

Albert Falco and Christian Bonnici circle a sperm whale in their inflatable boat.

manded only courage and physical effort. It was also the product of a mind accustomed to unusual marine situations, a mind totally involved in this adventure.

On land some men exercise a certain magnetism over animals. They can get close to wild beasts, tame them, and make them obey. This is what Falco did with sea animals, an undertaking much more difficult than on land. How does a diver tell what a whale is thinking behind its little blue eye? How can he tell what a shark will do, whether it will attack or flee? Such animals are much stronger in the water than man. They are in their element and it is impossible to escape if they become violent.

Still, as chief diver of the *Calypso,* Falco proved that sharks could be confronted. If men came to dominate them, it was through an understanding of their behavior. They know that if they become frenzied they cannot be controlled.

Falco's 40 years of experience taught him how to make friends of cetaceans or fish. No one could advise or guide him in this apprenticeship. He had to learn everything from his own sense of observation and from an innate empathy with the sea.

There is no doubt that his childhood at Sormiou Cove helped prepare him to be a diver, as did his spearfishing and explorations around Marseilles.

Spearfishing seems absurd and wrong to us today, but it must be admitted that it formed a whole generation of divers and contributed importantly to understanding the underwater world.

Familiar animals

After having killed many groupers during his childhood and adolescence, Falco began to make friends with them.

One became famous as "Jojo." It was one of the first brilliant actors of the film *World of Silence.* Like a Hollywood star, Jojo always wanted to upstage everyone else. It always wanted to be as close to the camera as possible, even if it had nothing to do in that sequence. It had to be shut in a shark cage to keep it out of the way.

Another grouper around the undersea houses was not always fortuitous, as Falco explains:

"We had put the houses on its territory. It tolerated us there, but wanted to show it was the landlord. There must have been some strange ideas running through its head. And to it we were strange-looking fish. Still, it did not fear to confront us. Once when Frédéric Dumas teased it, it bit him. Was it intelligent? One thing is certain: a grouper has consciousness and an individual personality.

"When we were filming turtles on Europa Island I encountered a strange grouper: it changed color. It would change suddenly from brown to white. It rubbed up against us. It lived on the south side of the island, about 45 feet down. It also took up with us, but it did not follow divers to the open sea, never leaving its well-defined territory."

The triggerfish

The triggerfish is almost as interesting as the grouper. It has a funny look with its big eyes and little mouth, and it does not appear very smart. But there is a ferocious will in that little body. Nothing can budge a mother triggerfish watching over its eggs and oxygenating them by fanning the water with its fins. It attacks all comers, including divers. It has surprisingly strong jaws, which can even tear the divers' neoprene suits.

"The one that Guilbert fed around the underwater house at Shab Rumi showed great initiative. It watched for each diver leaving the large house and followed him. But it also entered the entrance lock and capered around until Guilbert gave it something to eat. It realized that it would get nothing from divers on the bottom, so it hunted its own food. It moved aside bits of coral, even breaking them to find a crab or shellfish. It could break an oyster shell with its jaws and then gobble it, head down, its tail quivering.

"It also had its territory, even larger than the grouper's. When the divers went too far off, it left them and returned to stand guard beside the house."

Dolphins and liberty

Today we know that it is relatively easy to approach certain fish. Marine mammals are something else. To be sure, dolphins show a strange attraction to men and many stories attest to that. Still, relations between men and dolphins are

Diving at night off Yucatan, the men discover a giant sponge with their lights.

At left: At Loude Rock in Sormiou Cove, an underwater scene of gorgonia, coral, and ascidia.

not so simple, partly as a result of that mutual sympathy. It gives rise to tragic misunderstandings, and captivity does not always suit these animals.

Albert Falco was the first to understand the importance of the affective life of dolphins. He guessed at their psychic fragility and worked to lessen the shock of captivity. He attempted to comfort them physically by sharing their watery prison. He felt the same horror for those walls, for that pool in which they were suddenly trapped. He took the dolphin in his arms and led it around the pool, and it grew attached to him. It recognized him.

''That has left me some bad memories,'' he says. ''I'll never forget the dolphin that slipped out of my arms and killed itself by running into the wall. Since then I have never put another one in a pool. What we did on the *Calypso* was much less cruel. We kept dolphins only an hour or two in a pool formed by nets in the open water.''

Approaching whales

Groupers, dolphins, sea lions, and even sea elephants are animals on the human scale. There is no great problem in approaching them and even petting them. It's not the same with whales and sperm whales, however. These enormous masses seemingly swim slowly, but can actually attain 10 and even 18 knots in case of a chase.

A diver floating in the sea is nearly powerless. What could he do against a 30-foot whale weighing 30 to 40 tons? This impotence was well expressed by one of the first divers who tried to approach a sperm whale. When he returned to the *Calypso* he said, ''You might as well try to catch a locomotive!''

For years, the whole crew was fascinated by whales, trying to imagine ways to approach them. In the observation room on the *Calypso*'s bow they could watch the sperm whales swimming in a group. But the divers never tried to join those groups, not only because of the danger, but also because it would be impossible to catch up with animals going so much faster than they could.

''Yes,'' Falco says, ''that looks impossible unless something is invented. In our discussion we concluded that in any case it would be necessary to slow the animal from 18 knots to 2 or 3 to see it well in the water.''

This became necessary the day when it was decided to make a film about whales. To slow down a sperm whale from its 10-knot speed, divers pulled near in a boat and caught it with a small harpoon that did not penetrate its flesh but only the blubber. They left one to two thousand yards of nylon line to run to a float.

When the animal surfaced to breathe, it would offer the chance to swim alongside it or even accompany it on a dive.

Chases

This technique had nothing in common with those used by whalers. Their boats were powered by oars or sail, and they could not approach whales unless they surprised one or came upon an immobile herd. Today whales are much less numerous, and when the *Calypso* sighted one it had to try to catch it at full speed.

They could have tried to wind the whale during a long chase, but that is cruel and ineffective because the whale covers a wide area and only a very fast motor could keep up with it.

To film a whale the cameraman has to be at the right place when it surfaces. But how can you tell where that will be? Even with three boats at once, you nearly always get there late. At the beginning of the chase it will be 1,500 yards off to the right or left, and when you get there it has already dived again. But its breath becomes shorter with each dive, and it is possible to stay over it with the boat. Its shadow can be seen in the limpid water, and when it comes up it can be hit with a small harpoon that does not wound it. Still, the operation is dangerous in view of the disproportion between this mass of 30 or 40 tons of fat and bone and the little inflatable boat.

"I don't think we worried too much about that," Falco says. "We were excited by the chase and wanted to see the animal up close. I wanted to see the form of its head, it's look, its body, its mass. When you're involved in a chase at 18 knots, you want to impose your will on such an animal. You have to fight for hours, then start over, modifying the motors and your tactics. You have to be careful, or the beast can get away or you can accidentally kill it.

"What we wanted to do was to show the life of these immense animals in the water. The public can't imagine how supple and graceful these heavy masses are as they move. They move delicately, without any brutality or aggressiveness.

"I never thought of the danger, even though it was real, and there were a few accidents."

Double page following: a fin-back whale in the calm of the evening at Cape Guardafui (now Asir, Ras).

Among friends

Falco sometimes encountered whales without having to harpoon them even slightly.

"One evening near the Valdés Peninsula when I was in the rubber boat I saw a whale not far away. It disappeared, then suddenly its head emerged right in front of me, practically touching the boat. Was I lucky? Was it curious? Who knows? Whales are known to be curious, and I had the impression that this one looked at me, or at least at the boat.

"This one raised out of the water and then dived without leaving a ripple. It rubbed against my boat, but the boat hardly moved. It was like a dream. The seeming slowness of movement contributes to the dreamlike feeling. When you see those 35 feet of whale next to you, you say, 'It can't be true.' "

What happened around the Valdés Peninsula was also dreamlike. To the north, along the Argentinian coast, there is a sheltered bay where whales come to mate. The *Calypso* being anchored in the Gulf of San Jose, Falco and Bonnici took off in a boat with a camera.

"I see four whales who seem to be making love. We drift with the wind until we are in the middle of the four, whose heads are out of the water. They are so close I begin shooting even before entering the water. Bonnici follows me, and we find ourselves in a group of whales and a dozen dolphins playing around us."

It was the start of an extraordinary series of filming sessions. Every day Falco took the boat out, stopped the motor, and found the whales and dolphins, which soon numbered 30.

He dived next to a whale and looked at its eye before sliding alongside that great body, passing under and sometimes over it. The dolphins also came up close to the whale.

"I have never seen whales so easy to approach," Falco says. "I think they accepted the divers' presence because they took them for a species of dolphin."

Exploding boilers

The stay in the Gulf of San Jose was not always so peaceful. In mating season, the two or three hundred whales there pay little attention to men. Each female is surrounded by five or six males that throw themselves on her at full speed.

The jamboree of American submarines held off California during the making of the television film, *Those Incredible Diving Machines*.

All these giant beasts stir up a great deal of water, and the noise of their snorting is infernal.

"It sounds like exploding boilers," Falco says. "We would approach slowly in the boat, get into the water, and slip into the middle of this swarming activity. Suddenly one of the whales would swim right by and slap us with its tail. We were thrown right and left and bumped with heads, but no whale ever fled us. We could shoot all the film we wanted as long as the camera wasn't knocked into the air.

"We shot many sequences. Unfortunately, the water was not very clear and we had only 10 or 12 feet of visibility, not much when there's that much action.

"When the coupling was over, the whales disappeared right and left and the female generally was left by herself. We stayed with one that had several dozen dolphins around her. Slowly we slipped into the group of dolphins, and the whale let us come up to her. We stayed right alongside her for an hour as she swam very, very slowly. We swam next to her and ahead of her. She looked at us, and I filmed her eye, her jaw, her fins, and even the parasites on her skin.

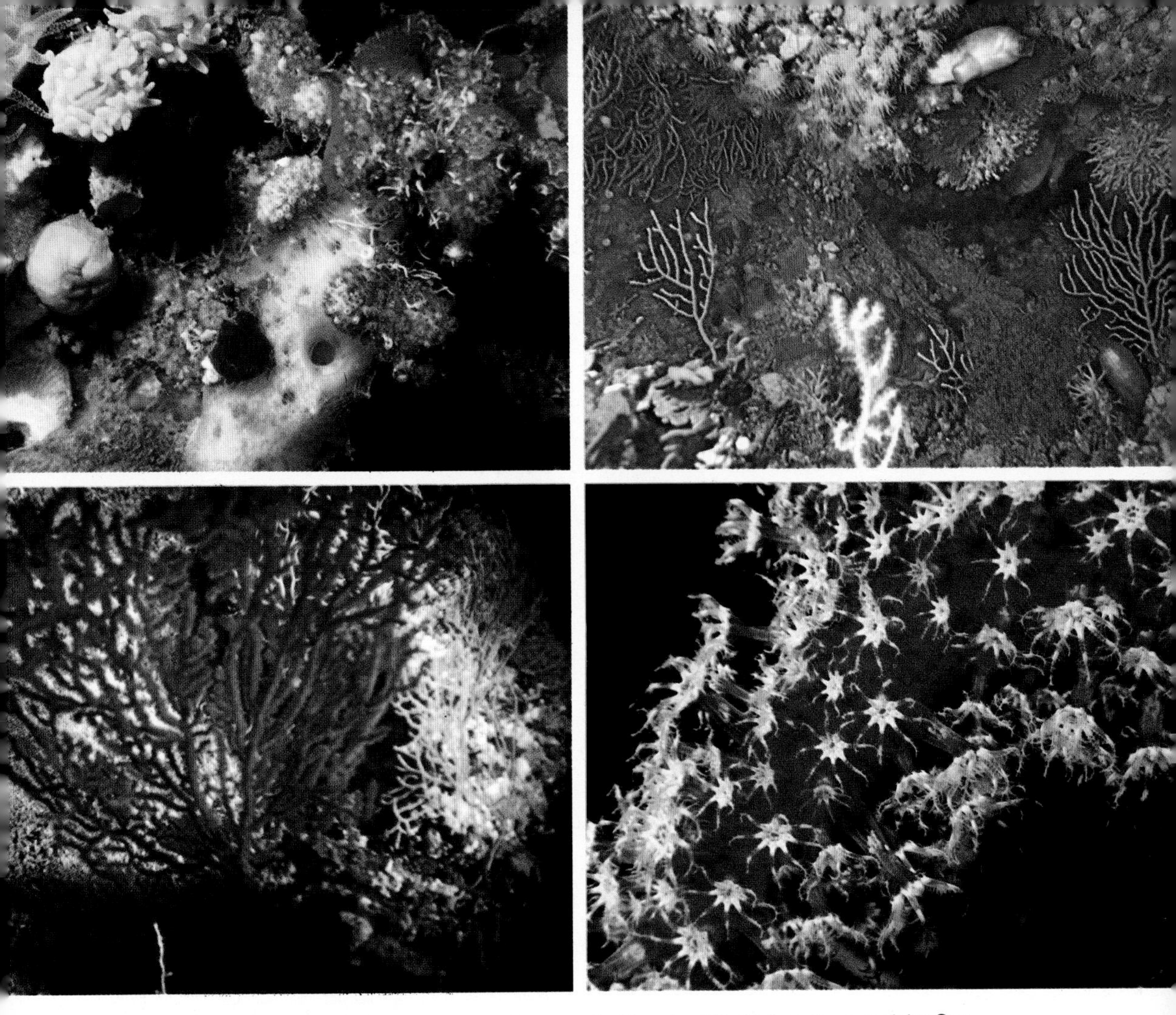

Views of the bottom of the Mediterranean: a rock covered with animal life 90 feet down; at right: Gorgonia, anemones, an ascidium. Lower left: A photo taken from the saucer at 590 feet showing gorgonia and a close-up of an Alcyonaria.

Falco feeding a triggerfish.

"Unfortunately the whale turned slightly and touched the outboard motor on our boat, lifting it with her tail and damaging the 40-horsepower motor. The spell was broken and she took fright and left. It was over. I would have liked to stay much longer with this beast that tolerated our presence so well. But at least we were able to stay together an hour."

That was Falco's most spectacular result. There is no doubt that thanks to him great progress has been made in learning to approach these mammals that were thought inaccessible. We still have a lot to learn from them.

Brief visits

This contact with whales, seals, fish—this revelation of life in the seas—is the great payoff for divers. And yet it does not satisfy them entirely. Those who have explored the underwater world for the last 25 or 30 years know that the scuba gives them access to only a very small part of it. Penetrating into the waters was long a dream of mankind, and when the underwater adventure began it appeared mavelous to us. But now we know that we have not gone very far. A diver with his scuba can visit only a tiny part of the continental shelf. Exploration remains limited to the edge of the shore and the immediate vicinity of islands. Above all, the depths man can attain do not go beyond 150, 200, 230 feet. This is a derisory figure compared to the average depth of the sea, which is 1,300 feet, or the depth of the Philippine Trench, which goes down to 34,578 feet.

We do not know what animals live at such depths. Nets have brought back a few small, monstrous-looking fish. We can only guess at the existence of giant cephalopods. One is said to have measured over 190 feet. Without reaching that size, the squid that live at great depth certainly measure 25 to 30 feet. And no one has seen the epic struggles between them and the sperm whales that feed on them. That is the life of the sea, and it evades us.

From 180 feet on down, sometimes before that, narcosis due to nitrogen endangers the diver, skewing his mental processes. When surfacing, the decompression pauses become so long that they make it impossible to stay at great depth more than a few minutes. Spending a few seconds at 220 feet is not really visiting the sea. Despite all the progress made over the last 25 years, it would be presumptuous to pretend that we know the sea well. Perhaps we know it less badly than before, but much remains to be discovered. For that we need the means.

We have seen that the danger of narcosis can be avoided through use of helium, but that helium is expensive. Such dives require a decompression bell and trained personnel. That is possible on the *Calypso,* but such equipment is not within the means of most people. Cousteau found another solution to narcosis with the diving saucer in which the underwater explorer breathes air at atmospheric pressure and can descend to 1,100 feet without danger.* But not everyone can afford such a vehicle. If Falco has been able to use one to learn more about the sea 1,000 feet down, it is thanks to making Cousteau's films and to his industrial and scientific assignments.

He has gone down with the saucer to check a pipeline, inspect a wellhead, or follow a canyon at the request of a geologist or biologist. He has always thought that these local incursions did not constitute a real knowledge of the sea, a systematic exploration of the bottom. He has often regretted that his official tasks kept him from satisfying his personal curiosity.

"Between Sicily and Tunisia," he says, "I was following a nylon line 750 feet down that marked the emplacement of a pipeline. Suddenly I sighted three large anchor stocks on the bottom, with the shipwreck just behind them. There was a pile of amphorae and pottery there, but I had to keep folowing the nylon. I left that ancient wreck behind, and it is probably lost forever."

---

*And even lower: *Sea Fleas* go to 1,600 feet, and the S.P. 3,000 to 10,000 feet.

A peacock fish in the Bahamas.

The Régalec

Falco dreamed of having his own submarine with which he could dive below 210 feet. He saw the answer while filming a large assemblage of diving craft at Catalina Island off California. They were of all sizes and shapes.

"That is where I discovered the *Nekton,* " Falco says, "a little submarine

The inflatable boat leaving the *Calypso* on a mission with Michel Deloire, Dominique Sumian, and Albert Falco aboard. At left, Commander Cousteau gives his last instructions.

Placing Albert Falco's small submarine, the *Régalec,* in the water during tests.

put together by an American garage mechanic. It is 10 feet long, with pointed ends and a turret in the middle. It holds two persons and goes at 3 knots for three or four hours. Its inventor let me try it, and it handled very easily. That was a good lesson for me.

''Returning to France, I got together a group of friends with all the necessary qualifications. We bought two old buoys that had supported antisub nets from a junk dealer, and put these two spheres together with a tunnel between. A large porthole was cut in the front sphere, and the rear one was topped with a Plexiglass turret giving 360-degree vision.

''It is propulsed by two screws on either side, and a propeller mounted horizontally makes it go up or down.

Albert Falco aboard his "Sunday submarine" able to dive to 3000 feet.

''A battery would have been too expensive and complicated to maintain, so power comes through a cable from an accompanying boat. The boat can be simply an inflatable one carrying a generator.''

The submarine was christened *Régalec,* from the name of a little fish that swims in a vertical position and is often found around cables or anchor chains. The *Régalec* has already made dives to 375 feet and has tested out well.

Double page following: a "fish wall" off Yucatan. It is a school of Haemulons, or grunts.

"I think that it will do everything I hope for," Falco says. "It will let me dive to 300 feet and look at everything I'm interested in without worrying about a work schedule. It will be my Sunday sub, and when I'm too old to dive I will still be able to cruise around in it. I believe the *Régalec* marks an important departure in underwater exploring because it has no utilitarian or scientific role. It serves simply to look around on the sea bottom, something future generations will do more than we have.

"I know just where I want to use it: along the French coast from Port Vendres to Monaco, where there is a plateau about 250 feet down. It's the one that leads to Rech Lacaze Duthiers and Cassidaigne Canyon, and I already know bits of it. I'm sure it will be rich in discoveries.

"It has considerable archaeological interest, for one thing. There are numerous wrecks off Narbonne, Marseilles, and Fos, not only Roman ships but Etruscan, Greek, and Punic ones. There are also airplanes near modern shipwrecks.

"Then there's the underwater scenery, about which we know little. We know the areas around a few cliffs and islands, but the sea is practically unexplored from 220 feet on. Even the middles of many coves are unexplored. Few people have been on the plateau I mentioned, where every rock is covered with Bryozoa, gorgonia, and sponges whose shapes and colors change with each species. And there are swarms of antedons and brittle stars.

"The idea is to acquire an intuitive feel for the sea that we lack at present. On land, the valleys, forests, and animals seem to be all logically ordered, but the structures of the sea are still unknown to us, and we don't understand the significance of what we do see. But bit by bit our 'sense of the sea' is developing. I can already predict that around a Mediterranean canyon I will find bushes of Antipatharia, chalice-shaped sponges on flat rocks, yellow or white coral on larger rocks, red coral and certain algae on overhangs, rockfish on ledges swept by currents, and so on."

This is the mesage of Albert Falco, the first leading underwater man of our time. Twenty-five years of diving in all the world's seas have given him a fund of experience without equal. Trained from childhood to block his breath to dive, used to moving in a silky-soft, viscous fluid, he has been shaped all his life by the sea. Perhaps prefiguring heroes of the future, he has responded to a consuming desire to know and understand a world foreign to our minds and senses. To his exceptional destiny he has brought a great humility, an endless admiration for life, instinctive reactions to unusual situations, an adaptability to a hostile environment where for thousands of years man dared not enter.

"One lifetime is not enough," Falco says. "There are thousands and thousands of things to observe. I won't be able to see it all, even if I limit myself to the area I have decided to concentrate on. I would need several lives. The sea is so large. The nineteenth century was the era of great explorations on land. Now the hour has come for the true discovery of the sea.''

Page following: Albert Falco returning from a dive.

appendices

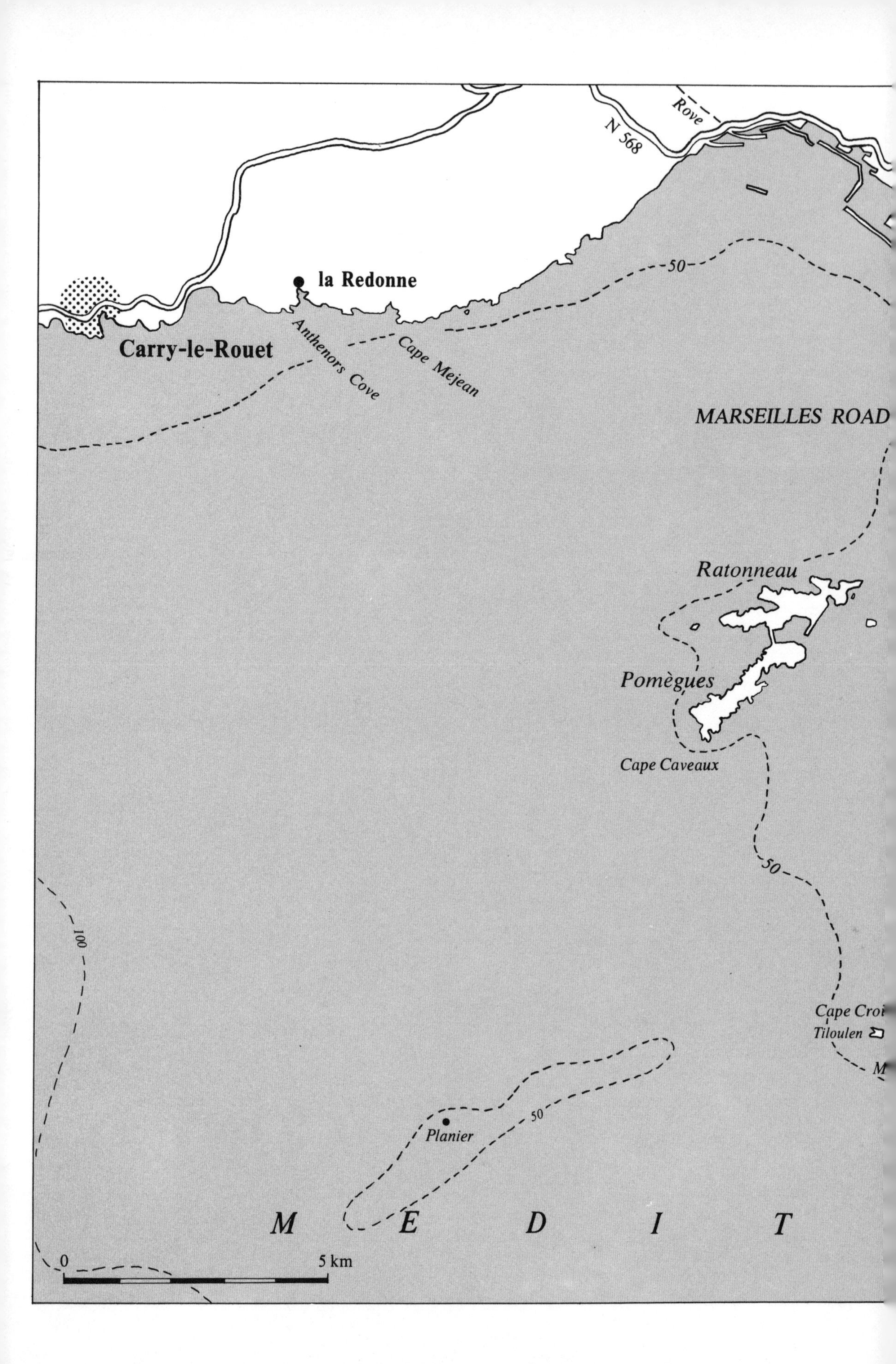

Rove
N 568
la Redonne
Carry-le-Rouet
Anthenors Cove
Cape Mejean
50
MARSEILLES ROAD
Ratonneau
Pomègues
Cape Caveaux
50
100
Cape Croi
Tiloulen
M
50
Planier
M E D I T
0
5 km

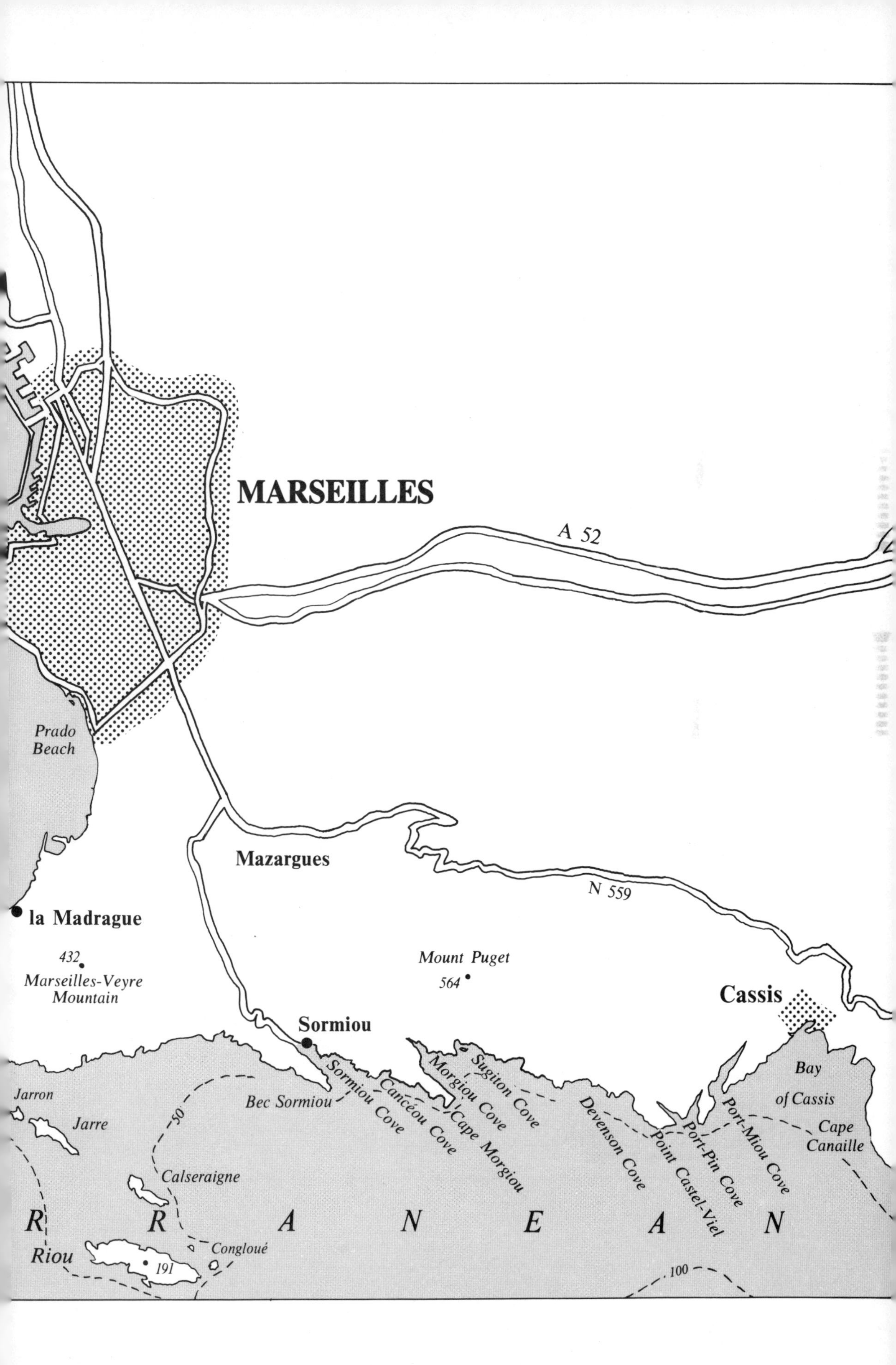
MARSEILLES
A 52
Prado
Beach
Mazargues
N 559
la Madrague
432
Marseilles-Veyre
Mountain
Mount Puget
564
Cassis
Sormiou
Sugiton Cove
Morgiou Cove
Bay
of Cassis
Jarron
Bec Sormiou
Sormiou Cove
Canceou Cove
Cape Morgiou
Devenson Cove
Point Castel-Viel
Port-Pin Cove
Port-Miou Cove
Cape
Canaille
Jarre
50
Calseraigne
R
R
A
N
E
A
N
Riou
191
Congloué
100

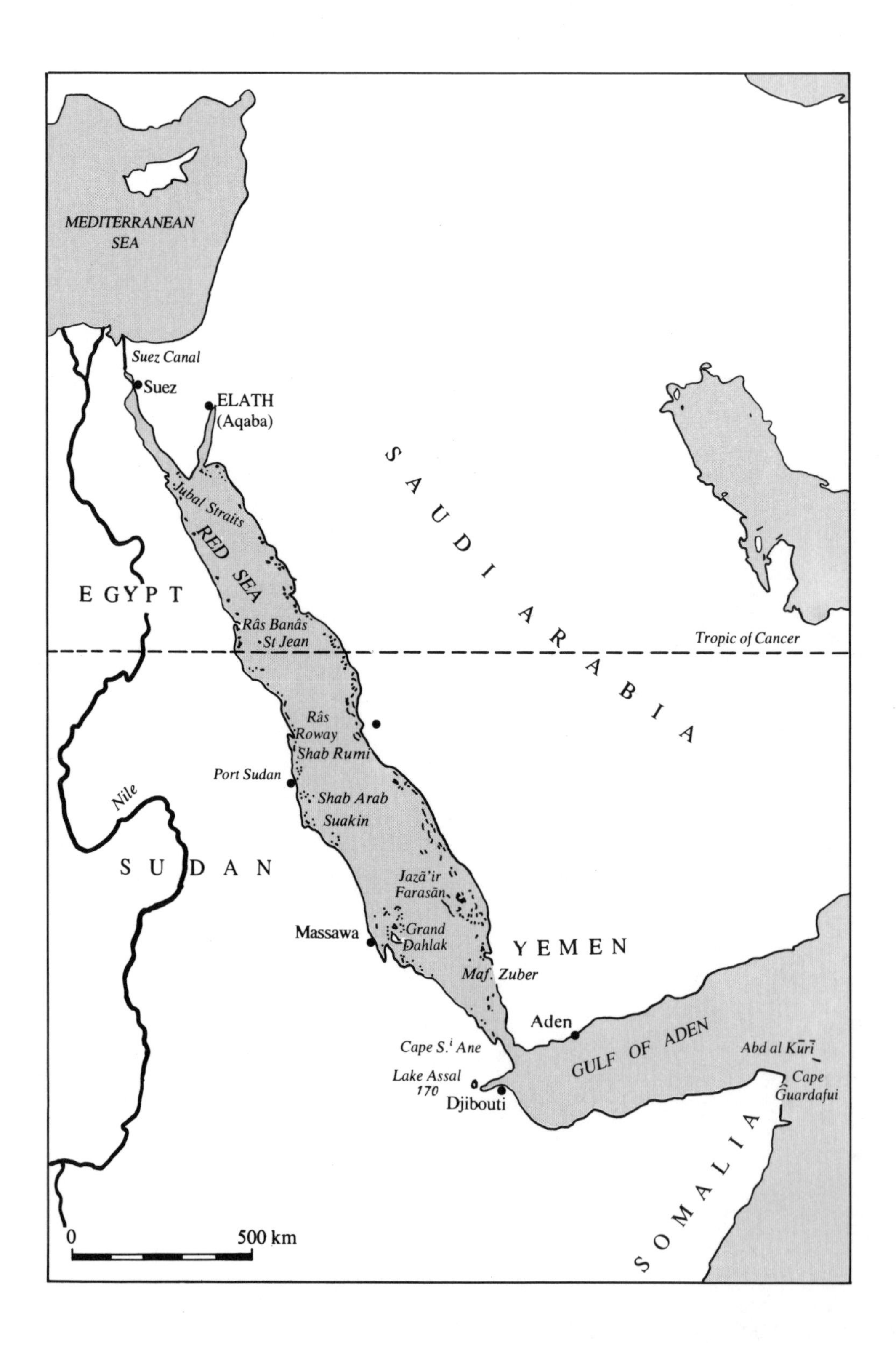
MEDITERRANEAN SEA
Suez Canal
Suez
ELATH (Aqaba)
SAUDI ARABIA
Jubal Straits
RED SEA
EGYPT
Râs Banâs
St Jean
Tropic of Cancer
Râs Roway
Shab Rumi
Port Sudan
Nile
Shab Arab
Suakin
SUDAN
Jazā'ir Farasān
Grand Dahlak
Massawa
YEMEN
Maf. Zuber
Aden
Cape S.i Ane
GULF OF ADEN
Abd al Kuri
Lake Assal 170
Djibouti
Cape Guardafui
SOMALIA
0
500 km

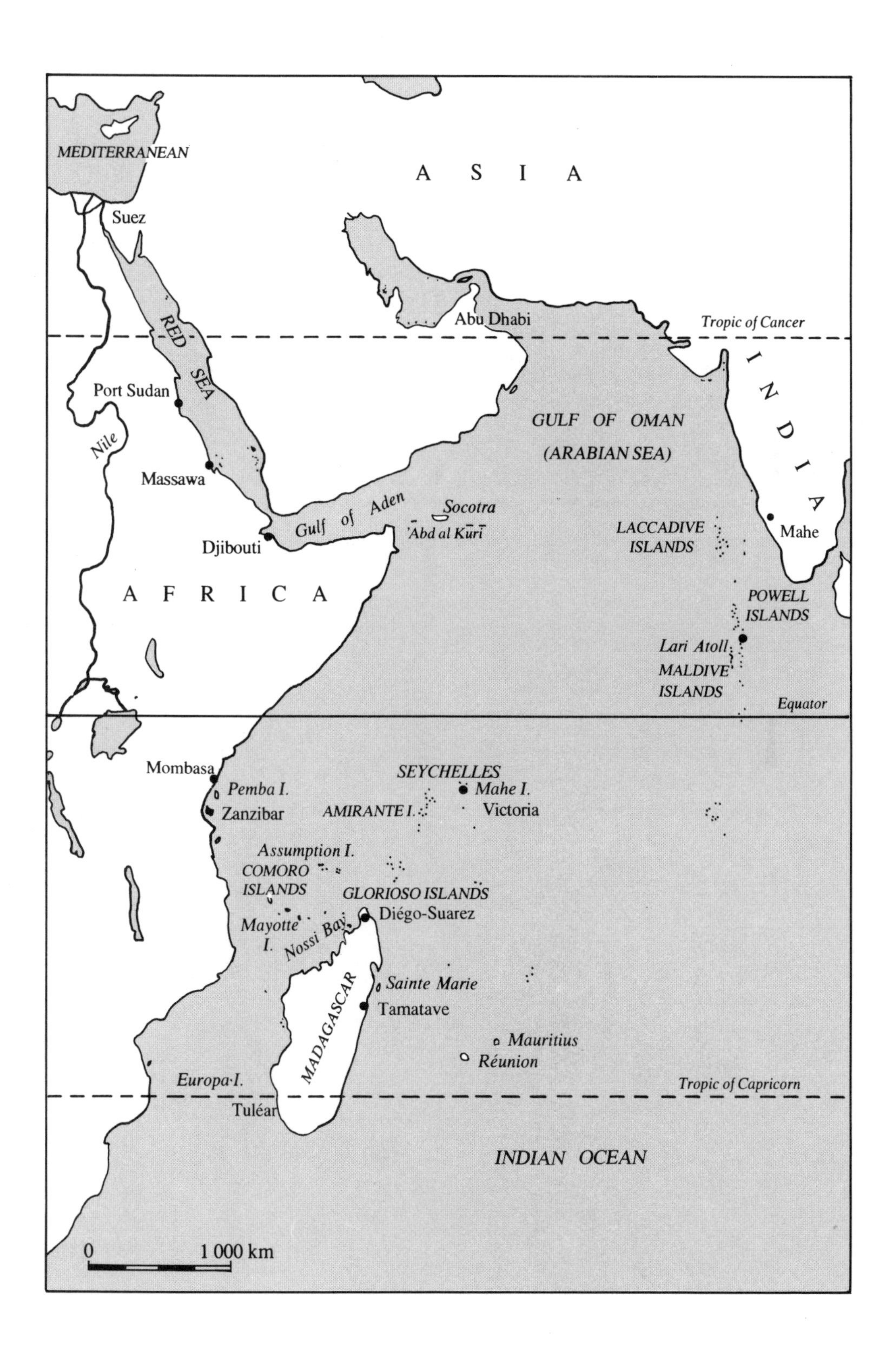

MEDITERRANEAN
A S I A
Suez
RED SEA
Abu Dhabi
Tropic of Cancer
INDIA
Port Sudan
Nile
GULF OF OMAN
(ARABIAN SEA)
Massawa
Gulf of Aden
Socotra
'Abd al Kūrī
LACCADIVE
ISLANDS
Mahe
Djibouti
A F R I C A
POWELL
ISLANDS
Lari Atoll
MALDIVE
ISLANDS
Equator
Mombasa
SEYCHELLES
Pemba I.
Mahe I.
Zanzibar
AMIRANTE I.
Victoria
Assumption I.
COMORO
ISLANDS
GLORIOSO ISLANDS
Diégo-Suarez
Mayotte
I.
Nossi Bay
Sainte Marie
MADAGASCAR
Tamatave
Mauritius
Réunion
Europa I.
Tropic of Capricorn
Tuléar
INDIAN OCEAN
0
1 000 km

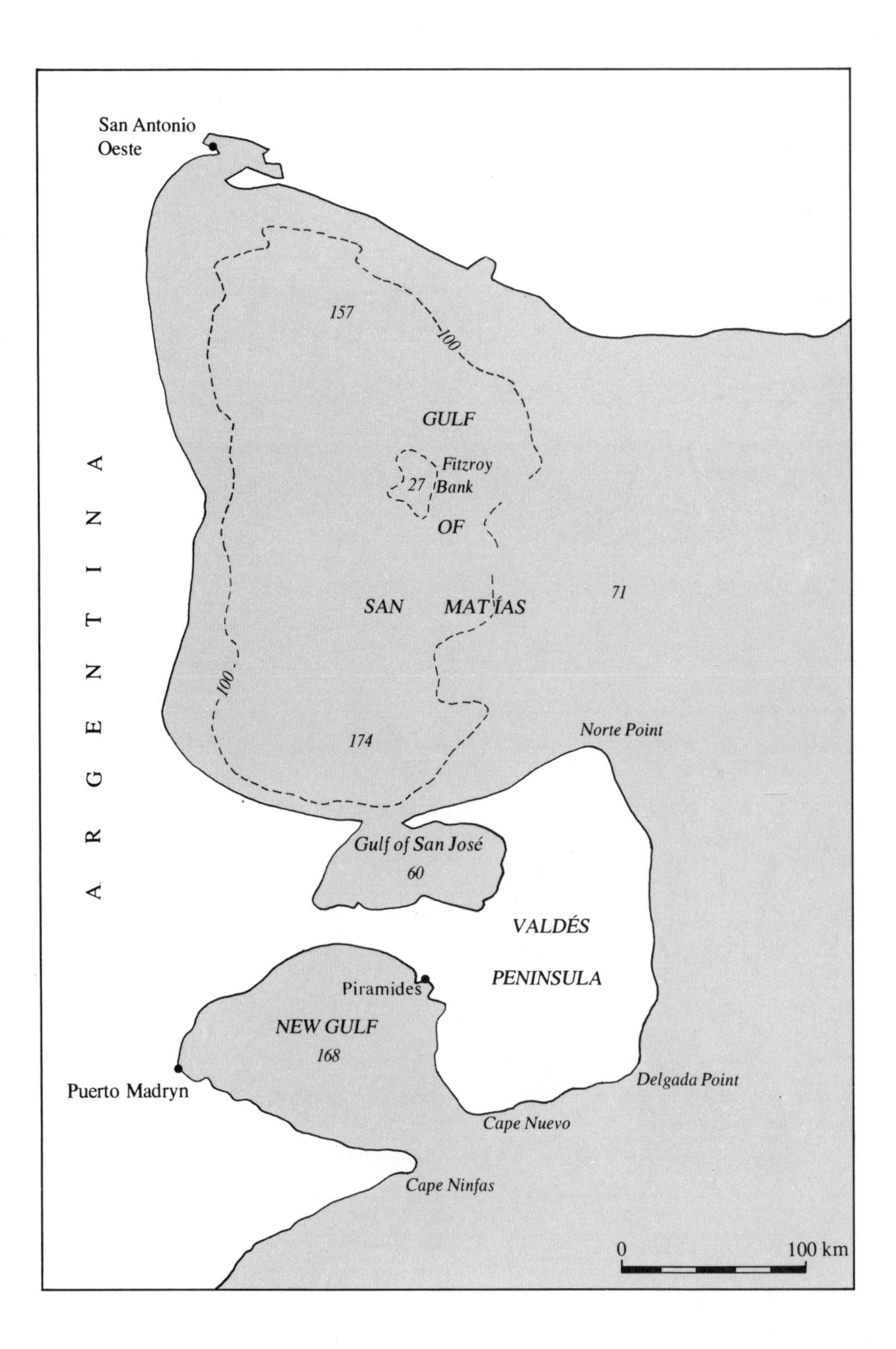
San Antonio
Oeste
157
100
GULF
Fitzroy
27
Bank
OF
A R G E N T I N A
SAN
MATÍAS
71
100
174
Norte Point
Gulf of San José
60
VALDÉS
PENINSULA
Piramides
NEW GULF
168
Puerto Madryn
Delgada Point
Cape Nuevo
Cape Ninfas
0
100 km

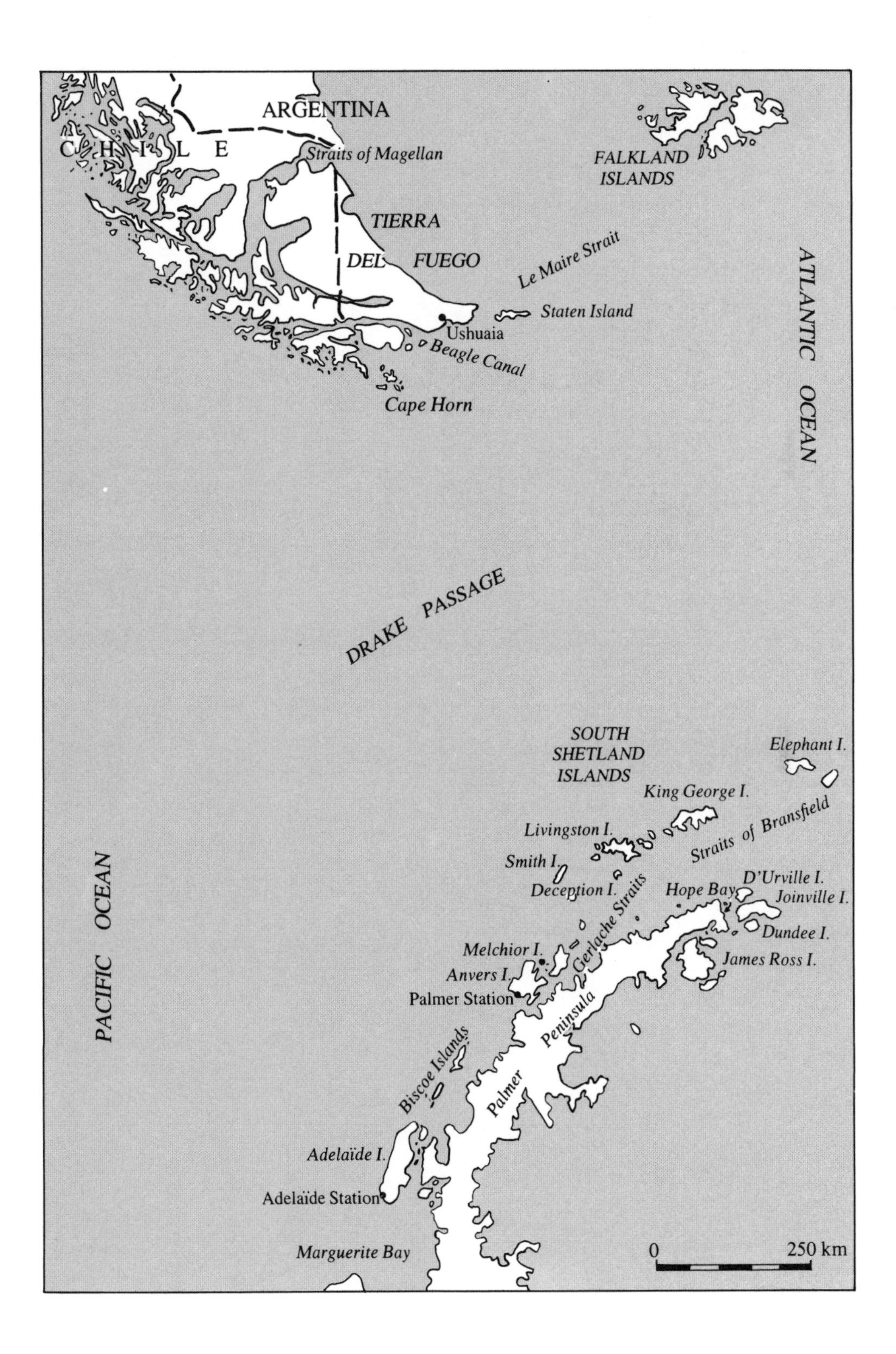

ARGENTINA
C H I L E
Straits of Magellan
FALKLAND ISLANDS
TIERRA DEL FUEGO
Le Maire Strait
ATLANTIC OCEAN
Staten Island
Ushuaia
Beagle Canal
Cape Horn
DRAKE PASSAGE
SOUTH SHETLAND ISLANDS
Elephant I.
King George I.
Livingston I.
Straits of Bransfield
Smith I.
Deception I.
Gerlache Straits
Hope Bay
D'Urville I.
Joinville I.
Dundee I.
James Ross I.
Melchior I.
Anvers I.
Palmer Station
PACIFIC OCEAN
Palmer Peninsula
Biscoe Islands
Adelaïde I.
Adelaïde Station
Marguerite Bay
0
250 km

appendix I

DIVING WITH THE SCUBA

For thousands of years man dreamed of being able to move with ease in the sea. Innumerable machines and devices were invented, from the inverted ''cauldrons'' of Aristotle to the diving suit with helmet and air tube of the nineteenth century.

In 1925 Commander Le Prieur conceived an autonomous diving device using compressed air, but it was not entirely automatic, since the flow of air to the mask had to be adjusted for depth.

The diving gear currently in use was conceived in 1943 by Commander Cousteau and an engineer, Emile Gagnan. It is a so-called open-circuit breathing apparatus that evacuates used air directly into the water. Also, it uses a ''demand system'' of air flow, providing air for each intake of breath but not continuously as in Le Prieur's gear.

The diver wears one or more bottles of high-pressure compressed air on his back. He clamps his teeth on a mouthpiece linked by two flexible tubes to a regulator on the bottle. The regulator delivers air at a pressure equal to that of the am-

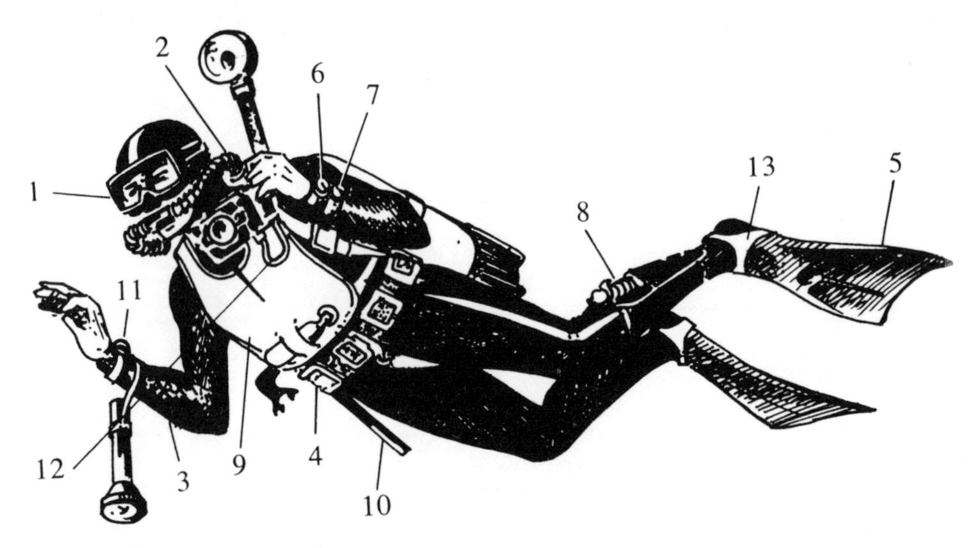

Cousteau-Gagnan scuba, *Mistral* model, and diving accessories.

INDISPENSABLE EQUIPMENT

1. Mask
2. Regulator and bottle of compressed air
3. Wet suit
4. Lead belt
5. Flippers

NECESSARY EQUIPMENT

6. Depth gauge
7. Watch
8. Dagger
9. Inflatable life vest
10. Snorkel

USEFUL EQUIPMENT

11. Compass
12. Camera and flash attachment
13. Heel straps to hold flippers

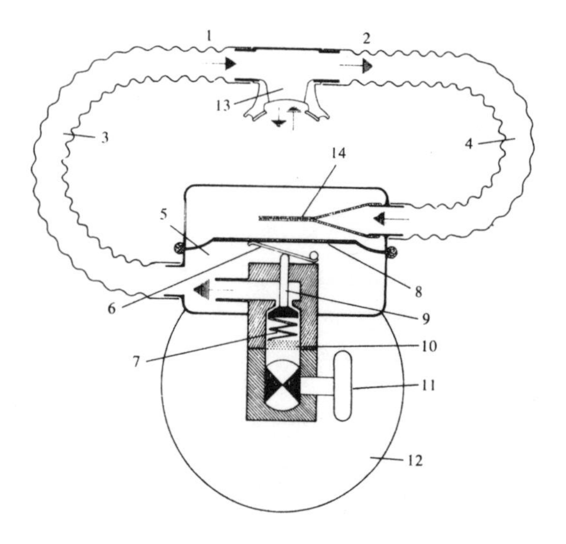

Diagram of Cousteau-Gagnan regulator.

1. Inhaled air
2. Exhaled air
3. Inhaling tube
4. Exhaling tube
5. Air at low pressure
6. Lever responding to demand
7. Spring
8. Membrane
9. Valve responding to demand
10. Filter
11. Stop-cock
12. Bottle of compressed air
13. Mouthpiece
14. Exhaust valve

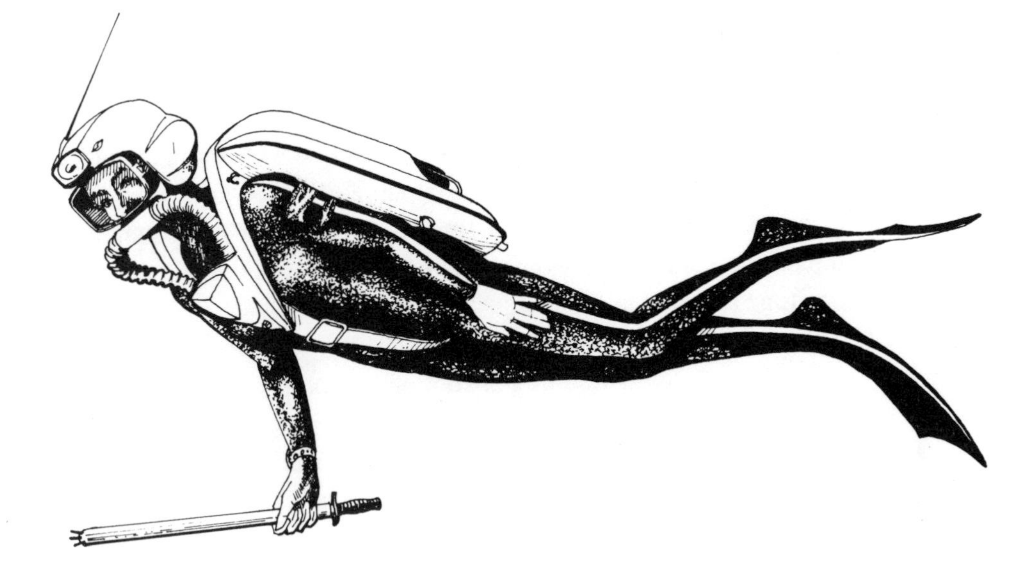

The latest model of scuba equipment, including a helmet light and a built-in telephone. The diver is holding a shark stick.

bient water. The used air returns to the regulator and is evacuated into the water through a "duck's bill," leaving a string of bubbles in the diver's wake.

This simple, reliable, entirely automatic device is easy to use, making scuba diving possible for a large public and facilitating underwater exploration. It constitutes a considerable improvement over the diving suit with helmet and air tube, which is more complicated, uncomfortable, and dangerous to use, as well as more difficult to master. The "lead foot" diving suit is tied to its air tube and line to the surface, and its radius of action is thus severely limited. With a scuba (*S*elf-*C*ontained *U*nderwater *B*reathing *A*pparatus), a diver moves freely in the water and is not prone to the diving accidents of the nineteenth century, particularly the possibilities of being crushed or of having the suit over-inflate and rise suddenly. On the other hand, he must always be aware of two other dangers that also threaten helmeted divers: rapture of the depths, or narcosis, and decompression accidents on surfacing.

Rapture of the depths is a narcosis that affects some persons at 120 feet, whereas others do not feel it until further down. It is related to the presence of nitrogen, and involves a dangerous alteration in the reasoning faculty. By replacing nitrogen in the breathing mixture with lighter gas such as helium, the threshold of narcosis is changed by several hundred feet.

Decompression accidents occur when nitrogen comes out of solution in the body and forms bubbles; the accident becomes more serious the faster the diver surfaces and the longer and deeper he has dived. Thus divers should observe the necessity of surfacing slowly to give the nitrogen the time to eliminate itself from the system. Tables have been established that indicate the number and length of pauses to make while surfacing, depending on depth reached and time spent there. A very short dive does not give the system time to absorb a dangerous quantity of nitrogen and pauses are then unnecessary. But when depth and duration increase, decompression time becomes considerable.

That is the reason for the experiments with "underwater houses." The human body becomes saturated with nitrogen after a few hours of high pressure and remains that way however long the diver remains at that pressure. It is therefore advantageous to decompress only once, after several days or even after a month, as during the Conshelf III experiment. Thus a long stay under water requires only one surfacing, and that is important because, as Paul Bert said, "You pay when you leave."

Great depth

Eliminating nitrogen from the air breathed by divers constitutes an important step forward. The oxygen-helium mixture avoids narcosis, as was seen in Albert Falco's accounts of dives. This mixture is used by practically all professional divers working below 190 feet.

When the depth reaches that figure and the dive is relatively long, the number and duration of the pauses grows excessive. For a dive of 45 minutes at 200 feet, the decompression time is 3 hours 36 minutes. One hour spent at 300 feet requires 7 hours of decompression.

Such pauses are obviously very unpleasant in the open sea. Submersible decompression chambers such as the Galeazzi diving bell used by the *Calypso* crew have improved surfacing conditions for deep dives.

The diving bell is a steel cylinder with a bottom panel opening to the sea. Divers enter the bell at 100 or 120 feet down, close the panel, and are brought to the surface while the pressure inside remains that at which they entered. Decompression can be effected during surfacing of the bell and on the surface itself as slowly as necessary and under medical supervision. Oxygen can be inhaled when pressure equals minus 40 feet.

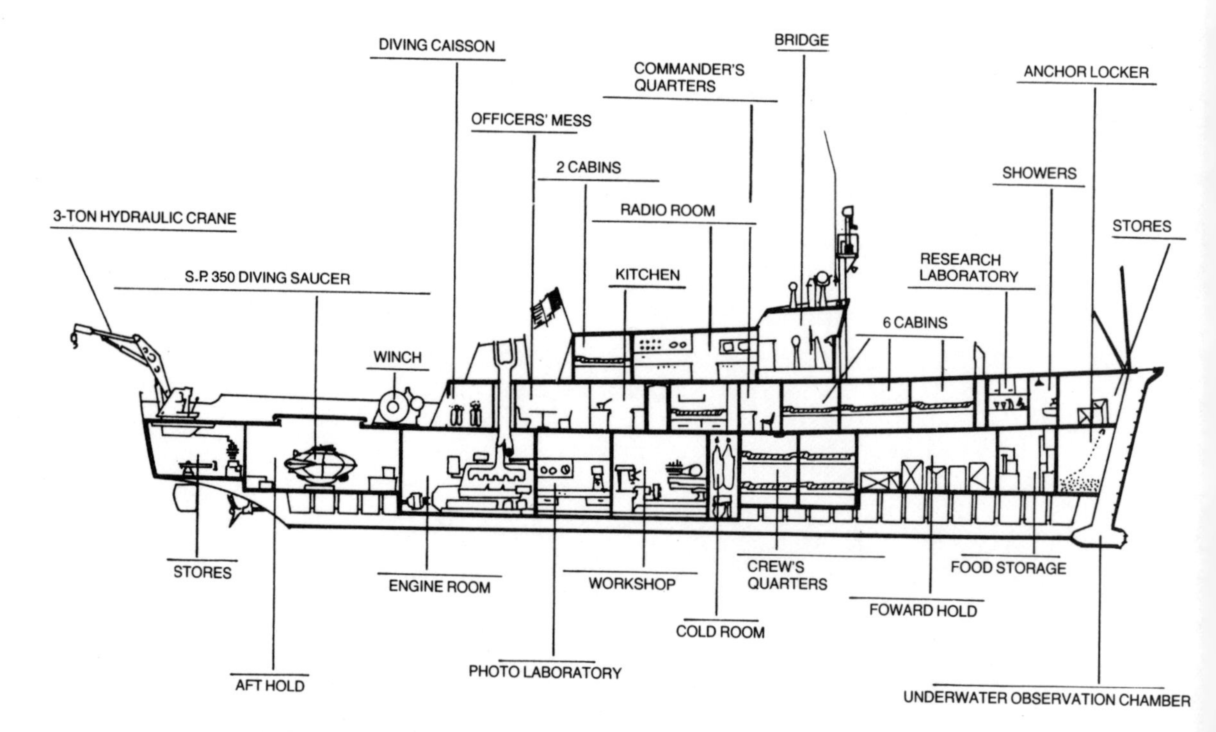

A former mine sweeper built in the United States, the *Calypso* has been modified several times to make it suitable for oceanographic research. The bridge and map room have been completely transformed. Cabins have been added forward. The aft hold serves as the diving saucer's garage.

appendix ii

THE *CALYPSO*

The oceanographic research ship the *Calypso* is a former mine sweeper with a double hull of wood that was constructed in the United States in 1942 for the British government. Commander Cousteau, wanting to explore the underwater scene with the equipment he had designed, found this ship for sale in Malta. It already bore the name *Calypso.* It seemed to fulfill his criteria for a ship devoted to oceanography to be used by divers. Loel Guinness, a British subject who loved the sea, helped Cousteau to purchase and outfit the ship.

Measuring 141 feet long, 23½ feet wide, and displacing 329 tons, the *Calypso* is solidly built and exceptionally maneuverable. Its shallow draft, its two engines, and its two screws enable it to maneuver well in areas of coral reefs or ice floes.

Several alterations were made to turn the ship into J.Y. Cousteau's idea of a scientific research vessel. The first was an observation chamber installed in the bow—a metal well leading down to a turret with five portholes eight feet below the waterline.

A double mast leads to a small elevated platform on the foredeck. This carries the radar scope and is a lookout post for guiding the ship through difficult passages or for spotting marine animals like whales or dolphins.

About thirty persons live aboard, but room is limited due to the considerable amount of equipment necessary for underwater exploration.

The *Calypso* carries 20-odd scubas, underwater "scooters," "wet submarines," "diving saucers," two launches, two fast inflatable boats, and cinema equipment including cameras, floodlamps, and electrical cables. There are also several laboratories and aquariums aboard, including one that has anti-roll devices.

Everything happening on board and in the water can be followed on closed-circuit television. Communications among the miniature submarines, the divers, and the *Calypso* are handled over ultrasonic telephones. Tape recorders and hydrophones record sounds made by animals in the water.

Besides the usual modern navigational aids, the ship has a special sounder for great depths.

Thus transformed, the *Calypso* was the only French oceanographic vessel from 1951 to 1965. Its size and maneuverability make it particularly suited to that role. For 11 years it was used by researchers from all over the world, leading to the capture and identification of numerous species and to the publication of important scientific articles.

Cruises

The *Calypso*'s research cruises are not financed by any public or private subsidy. The Monaco Museum affords it only scientific help. The *Calypso* is managed by a foundation—French Oceanographic Campaigns—created by Commander Cousteau in 1950. It is financed solely through the Commander's royalties from books and television, and industrial and scientific contracts.

The *Calypso* made its maiden voyage in 1951 in the Red Sea with a team of biologists and zoologists aboard. Then at the Grand Congloué islet off Marseilles it carried out archaeological digs at a shipwreck dating from the Third Century B.C. It plied the Indian and the Atlantic oceans making films, notably *World of Silence* and *World without Sun,* and to effect scientific research such as photographing several deep trenches with the cooperation of Professor Harold Edgerton of the Massachusetts Institute of Technology

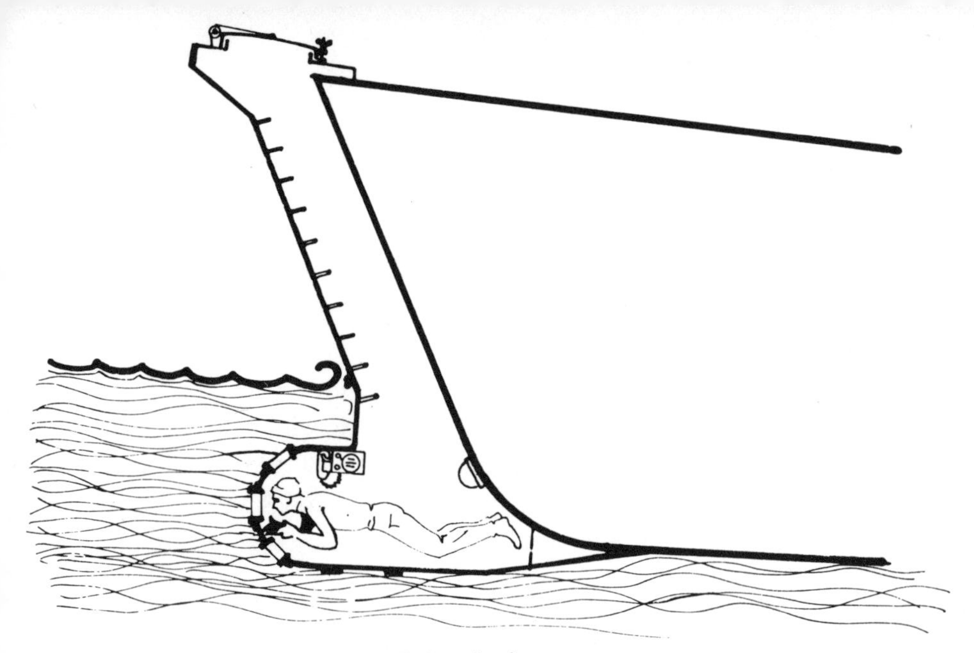

A metal stairwell descends to the *Calypso*'s observation chamber just fore of the bow. The chamber has five portholes for shooting stills and movies eight feet below the surface.

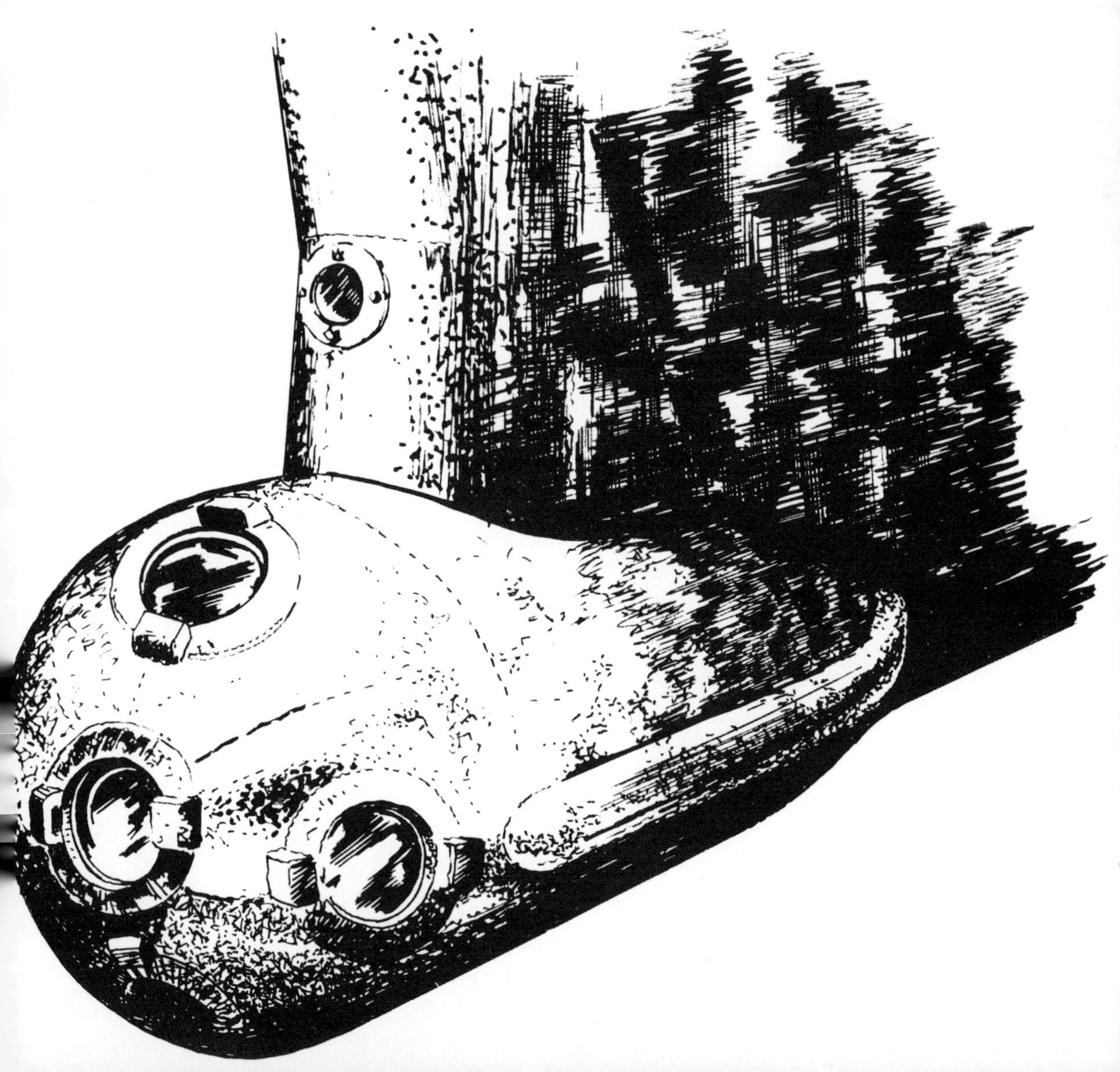

The *Calypso*'s longest cruise lasted three years, from 1967 to 1971. It was a voyage of 140,000 nautical miles across the Mediterranean, the Red Sea, the Indian Ocean, the Atlantic, and the Pacific to the Bering Straits. During this trip it made 24 films for world television.

After a stay in Marseilles interrupted by other films in the Mediterranean, the *Calypso* sailed on September 29, 1972, for a six-month expedition in Antarctica with stops and missions in Argentina, Patagonia, and the Falkland Islands. The object of this expedition was to study the effects of abusive hunting and the chemical intoxication of vulnerable warm-blooded marine animals such as whales, killer whales, seals, and penguins.

For this cruise the *Calypso* had special equipment for receiving satellite weather data from NASA. It also had a Hughes 300C helicopter that could be dismantled for storage in the hold. A pad for it was constructed over the bow. The *Calypso* also had a hot-air balloon aboard, equipment for cold-water diving, new cameras, and new underwater lighting.

During the winter of 1975–1976, the *Calypso* and its crew did underwater archeological research off Santorin Island with the support of the Greek government.

appendix iii

FROM BUSHNELL'S *TURTLE* TO COUSTEAU'S DIVING SAUCER

The American David Bushnell (1742–1826), a student at Yale, devoted his time and money to studying underwater navigation. He was the inventor of the first functioning underwater torpedo boat.

His craft was made of wood and was in the shape of two turtle shells joined and immersed vertically. The one-man crew turned one propeller by hand for forward and reverse movement and another for descending and ascending.

It is notable that Bushnell perfected all the elements of a submarine still used today: screws, water ballast, pumps, lead ballast, torpedo, depth gauge, valves, and so forth.

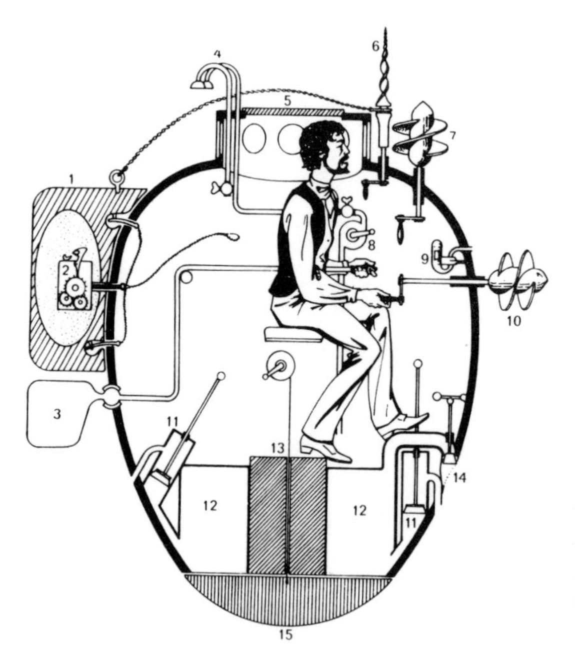

Bushnell's *Turtle* submarine

1. torpedo *2.* clockwork timer to set off the torpedo *3.* rudder *4.* ventilation tubes *5.* entry hatch *6.* drill for fixing the torpedo against a ship's hull *7.* vertical propeller for diving *8.* ventilator *9.* water barometer to indicate depth *10.* propeller for forward motion *11.* pump for emptying reservoirs *12.* reservoirs for water ballast *13.* ballast *14.* valve for admitting water *15.* safety ballast

In 1776, during the War of Independence, the *Turtle* got close to a 50-cannon British warship, but its pilot was unable to attach the explosive charge to it that would have blown it up. Regardless of this failure, Bushnell is considered the father of the submarine.

Commander Cousteau's diving saucer resembles a turtle floating horizontally and not vertically like Bushnell's craft. Its design principles, functioning, and testing have been sufficiently described in this book. However, we should note that besides the S.P. 350, which can accommodate two persons, there exist other similar craft designed by Commander Cousteau and produced by the Center for Advanced Marine Studies in Marseilles:

—The S.P. 1000, or *Sea Flea.* A one-seater designed to be used in pairs. With its claw it can assist or pull another *Sea Flea,* as has been demonstrated. It takes two exterior remote-controlled 16 or 35 mm cameras, along with hydrophones to record underwater sounds, particularly the cries of cetaceans.

—The S.P. 4000 or *Deepstar,* which can dive to 4,000 feet. It was designed for the Westinghouse Company and launched in 1966. It takes two passengers and cruises at 3 knots. It has made over 500 dives.

—The S.P. 3000, constructed for the French company CNEXO. It cruises at 3 knots and takes three passengers down to 9,850 feet.

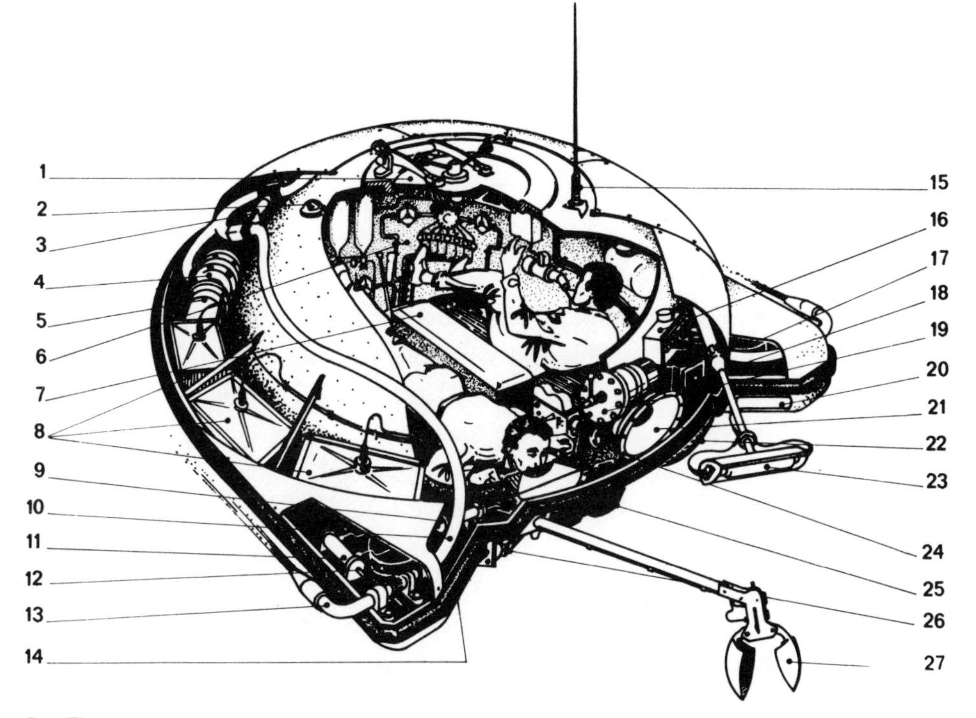

Sea Flea

1. Entrance lock *2.* Wide-angle portholes *3.* Fork to send water to right or left jets *4.* Pump *5.* Electric motor *6.* Interior fittings *7.* Water ballast *8.* Immersed batteries *9.* Control knob for claw *10.* Jet tube *11.* Control knob for fore or aft jets *12.* Transducer *13.* Jets *14.* Headlamp *15.* Radio antenna *16.* Electrical switch box *17.* Upper transducer *18.* Forward transducer *19.* Forward mercury balancer *20.* Flash *21.* Vision porthole *22.* Sample basket *23.* Telescoping light *24.* Camera porthole *25.* Releasable weights *26.* Stereo camera *27.* Claw for gathering samples

In the United States

Americans have generally used the classic cigar shape for their submarines rather than the elliptical shape of the diving saucers. This is true of the *Nekton* designed by General Oceanographics and capable of transporting two passengers and diving to 985 feet, with a range of 10 miles .

Another small American submarine, the *Alvin,* built by a division of General Mills for the Woods Hole Oceanographic Institute, has a slightly different shape due to its spherical cabin, but its diving capabilities are much greater: it descends to 6,500 feet with three passengers, its speed is 2 knots, and its autonomy seven to eight hours.

Like the *Nekton,* the *Alvin* is powered by a rear screw. Two small lateral propellers enable it to turn on itself or to accelerate diving or surfacing. The propulsion screw is also directionable.

One peculiarity of this submarine is that it can lighten itself by transferring oil into plastic balloons. The *Alvin* was used to recover the nuclear bomb dropped off Palomares, Spain, in February 1966. The bomb had been spotted by a remote-controlled craft, the *Curv.*

The *Alvin* is launched from between the two hulls of a catamaran. During one dive the *Alvin* broke its cable and sank. Its two passangers escaped through the open hatch, but the craft sank to 6,000 feet. It was recovered in 1969 by another, larger American submarine, the *Aluminaut.*

The General Dynamics Company has built a small submarine, *Star I,* which can dive to only 180 feet with one person aboard. It uses this craft to experiment with fuel cells, which could provide greater quantites of energy and eventually be the ideal solution for miniature submarines.

The model constructed in 1964, *Star II,* dives to 1,200 feet with two persons aboard and is powered by two engines, with a third for vertical movement. Also known as the *Asherah,* it has made 250 dives to date (used notably for archeological research in the Mediterranean).

A third model, *Star III,* presently in service in England, dives to 1,800 feet with three passengers. Like the previous American models, the *Stars* are propulsed by screws.

In 1964 the U.S. navy built the prototype of a light submarine in the shape of a sphere and named it *Deep Jeep.* It has no portholes, but features a viewing system composed of optical fibers. It carries two men. It dives to 1,800 feet and has two engines.

Diver vehicles

To assist divers in their work or to get to underwater houses, diver vehicles composed of two parts have been constructed. One part houses the pilot and the other one or more divers.

The first submarine of this type was constructed in 1967 in the United States by Perry Submarine Company, under the name of *Deep Diver.* From it was derived *Shelf Diver* in 1969, which was rented to the French firm Cocean. Weighing nine tons, this craft can take a pilot and three passengers to minus 820 feet. It has four to six hours' autonomy at two knots and 12 hours at one knot. It is composed of two spheres, the one in front for the pilot and the other for divers who leave through a lock using a narghile-type tube.

North American Rockwell Corp. constructed another such craft in 1967, the *Beaver IV.* With two spheres linked by a tunnel, the *Beaver IV* can drop off divers through the bottom of the rear sphere, or attach itself to the top of an underwater house to take on its inhabitants.

Launched from land on inclined rails, *Beaver IV* is powered by three motors and transports three men. It can reach a depth of 4,000 feet and has two powerful exterior arms.

Cable vehicles

A simple form of underwater craft is the type linked permanently to an accompanying ship by a cable. The cable does not support it, but simply supplies it with electricity. The submarine can thus draw on a much greater source of energy than can an autonomous craft.

Japanese adopted this cable system beginning in 1960 for their *Kuroshio I* and *Kuroshio II.* The latter can carry up to six passengers at minus 600 feet, and is linked to a ship by a 1,800-foot cable.

The United States has a cable craft, the *Guppy,* made by Solar, as well as an underwater observatory, *Nemo,* which is a sphere of thick plastic.

glossary

Acropora pharaonis

Colonial madrepore with numerous delicate branches that lives in calm waters.

Acropora hebes

A colonial madrepore with thicker branches than the previous species, very common in the Red Sea.

ACROPORA

Colonial Scleractinary, yellow, brown, green, violet, or blue, developing in the form of large branches or umbrellas. A very widely spread madreporaria in tropical seas, with multiple forms. In the Red Sea the umbrella form is especially well developed.

ALCYONARIA

An order of the class Anthozoa, phylum Cnidaria, subclass Octocorallia. The Alcyonaria are colonial and the tentacles and interior walls number eight.

The genus *Alcyonium,* that divers call "soft coral," has no skeleton but has microscopic calcareous spikes. Its colors are often beautiful: pink, green, bluish. Many swell during the night and attain considerable development. In Nouméa, New Caledonia, Dr. Catala observed a *Spongodes merletti* in an aquarium that went from a height of 3 inches to 16 inches in a few hours. There is then a marvelous, delicately colored, transparent arborescence.

ANGLER FISH

A member of the Lophiidae family, order Lophiiformes, class Teleostomi. The common Angler Fish *(Lophius pescatorius)* has a flat head that can be twice the size of the rest of the body. The mouth is enormous, with pointed teeth and powerful muscles. It is a "fishing fish" that hides in the mud with its dorsal fin sticking out with a lobe of flesh on it that acts as bait.

ANTEDON

A genus of Comatulids, class Crinoidea, phylum Echinodermata. The *Antedon* genus of the Comatulids (see individual glossary entry) includes many species of lively colors such as red or yellow, or more pastel colors. Comatulids group together on rocky bottoms where they attach themselves to rocks with appendages. They are composed of a disc with ten arms with which they swim.

ANTIPATHARIA

An order of class Anthozoa, phylum Cnidaria, subclass Geriantipatharia. Commonly called "black coral." Its colonies slightly resemble those of *Gorgonia*, but the polyps are different. The skeleton is very hard.

ASCIDIAN
Ciona intestinalis

1. Mouth (for ingesting food and oxygen)
2. Cloaca (for expulsing waste and reproductive elements)
3. Bud
4. Stolons
5. Muscles
6. Branchial sac (filter for food and oxygen)
7. Cerebral ganglion
8. Genital canal (male and female openings)
9. Intestine
10. Anus
11. Ovaries
12. Crown of tentacles
13. Stomach
14. Tunic
15. Derm

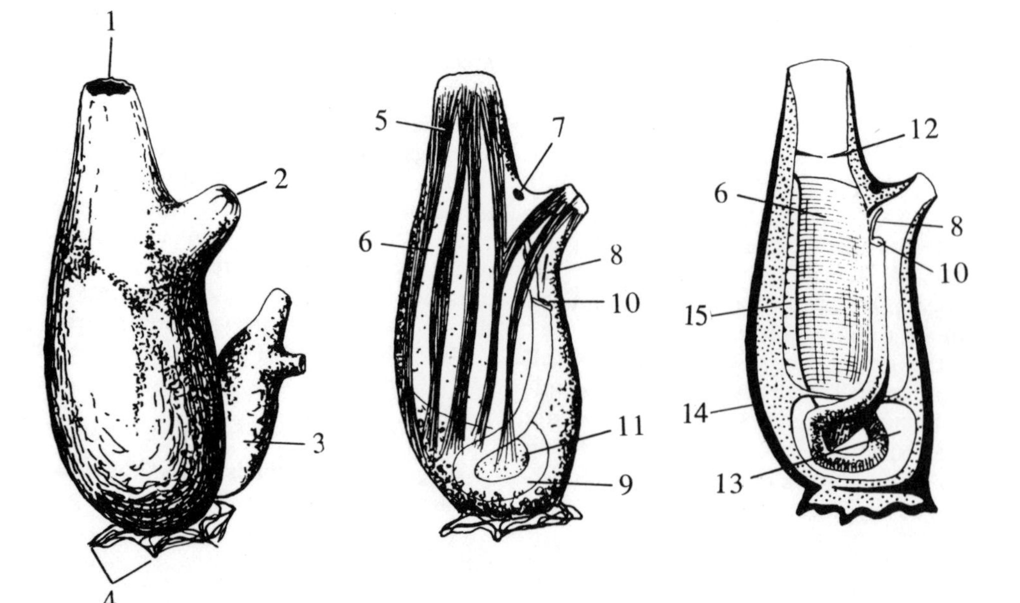

APNEA

Suspension of breathing for varied durations.

ASCIDIAN (SEA SQUIRT)

A member of the class Ascidiacea, phylum Chordata, subphylum Urochorda or Tunicata. It is a small Sea Squirt, sometimes bright red or yellow, that only attaches itself when fully grown; its larva is floating. This little sac has two openings, the forward mouth for entry of its sea water with food and a dorsal cloacal vent that expells waste. Despite its primitive look, the Ascidian possesses gills, a stomach, an intestine, and a V-shaped heart whose contractions push the blood 80 times in one direction and 40 times in the other. The Ascidian is hermaphroditic, having both a testicle and an ovary. The "violets" eaten on the Cote d'Azur are Ascidians.

BALLAN WRASSE

A common fish in the Mediterranean (family Labridae, order Perciformes), where many varieties exist. Its color varies from dark green to red-orange.

BARRACUDA

A carnivorous sea pike (genus *Sphyraenidae*) with visible teeth, large jaws, and a long body that looks like polished steel. The larger species can exceed six feet, and they swim in groups of three or four. The smaller ones move in schools whose members are all the same size and of the same age.

Barracudas have a bad reputation. In some regions they are said to be worse than sharks. They owe this unfortunate renown to their ferocious appearance, to their small hard eye, to their visible teeth, and to their behavior. They often stay obstinately with a diver, and if he threatens them they retreat briefly and then return. They are more spectacular than dangerous, and they seem not to live up to their frightening reputation. These handsome fish are found in all tropical seas, but the ones seen are often local species.

BETTE

Small pleasure and fishing boat of the Marseilles area of France.

BARRACUDA

BRITTLE STAR or SERPENT STAR

Ophiurida is an order of subclass Aphiuroidea, class Stelleroidia, subphylum Asterozoa, phylum Echinodermata. The body has the shape of a disc with five arms. The arms are thin and long and move rapidly. They can lift the animal and move it fairly fast.

On the ventral side of the disc there is a mouth, which corresponds to a digestive tube, a large stomach, an intestine. The Brittle Star has no anus. It lays eggs from which pelagic larvae hatch. It has been found in all seas down to a depth of 13,000 feet.

Brittle Stars can regenerate much of their body. They can spontaneously drop off one or several of their arms and replace them. They can also cut themselves in two and regenerate a whole body.

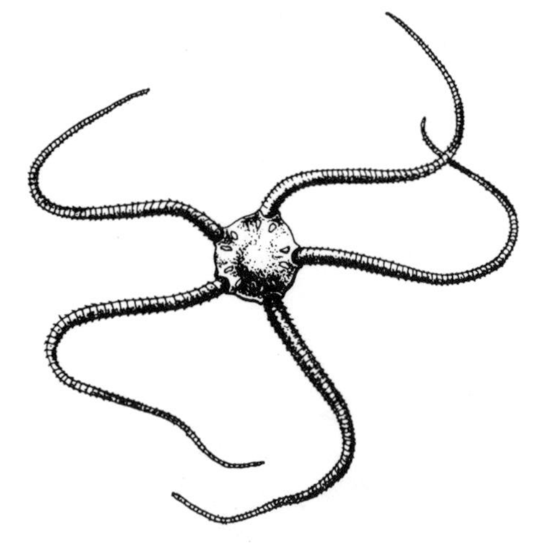

BRITTLE STAR

BROWN CAT SHARK

Small shark that frequents the Mediterranean coast. Both the small one, *Scyliorhinus caniculus*, and the large, *Scyliorhinus stellaris*, are oviparous; that is, they produce eggs that hatch outside the body.

CAT-TACKLE

The cat is the pulley that hoists the anchor.

CERIANTHUS or TUBE ANEMONE

Cerianthus membaranaceus of the class Anthozoa, phylum Cnidaria, with a shape similar to sea anemones. The mouth is surrounded by a supplementary row of small tentacles. The Tube Anemone is found in the Mediterranean on soft mud bottoms.

COMATULID

The Comatulid family belongs to the class Crinoidea, which themselves belong to the phylum Echinodermata. The Comatulid is composed of a very small disc with ten arms sticking out from it in pairs. The sexes are separate and the eggs are fertilized in the sea.

The Comatulids include numerous species with beautiful colors such as red, yellow, and orange. They can swim with their arms, but they seldom move from the corals or rocks where they are attached. They feed on very small animals which their minuscule tentacles bring to their mouths. The Crinoids were abundant during the Paleozoic era. Only a small number are left.

CONSTANT VOLUME SUIT

A watertight diving suit designed by Commander Cousteau that the diver can inflate with air from his bottles. Valves automatically evacuate excess air.

FEATHER-DUSTER WORM

EQUI-PRESSURE

Equal pressure on both sides of a wall or panel. The door of an immersed caisson can only be opened if the air pressure inside is equal to or superior to the water pressure.

FEATHER-DUSTER WORM or *SPIROGRAPHIS*

A member of class Polychaeta, phylum Annelida. It is a sedentary worm lodged in a tube from which it sticks out a colored plume that serves as a gill, also creating a current of alimentary water. In case of danger, the animal rapidly retracts the plume with a spiral movement.

GALATHEA

A Decapodan crustacean, suborder Anomura. It inhabits the coastal zone, generally under rocks. Two species are known on the French coast, *Galathea strigosa* and *Galathea squamifera,* measuring about 2 inches.

GIANT CLAM

The largest known bivalve mollusk *(Tridacna gigas).* It lives in the area of coral reefs in tropical seas. It can weigh over 400 pounds, and its valve can measure three feet long.

GORGONIA

An order of subclass Octocoralliaries, class Anthozoa, phylum Cnidaria. These ramified branches, or fans, can be yellow, violet, or pink and are in fact animal colonies. Their numerous polyps are distributed over a calcareous or corneocalcareous skeleton that is supple, not brittle. Branches of *Gorgonia* are attached to the bottom or to rocks, sometimes very tightly bunched. They adhere to the support with an encrusting base.

As with corals, the polyps can blossom out to the exterior or retract into a sort of calyx. They are found in all seas. In many tropical regions *Gorgonia* over three feet tall are an important part of underwater life. Their abundance and diversity was revealed by divers using scuba gear. Dives with the saucer revealed that in the Red Sea they virtually amount to forests at great depth.

GROUPER, also GIANT SEA BASS, SEA PERCH, or CALIFORNIA JEWFISH

A sedentary fish that lives in a grotto or hole in coral, preferably on a sandy bottom and at different depths. It attacks its prey with extraordinary speed, sometimes even "inhaling" it from a distance.

Formerly very abundant on the Mediterranean coasts where it was over-fished, it is also abundant along the coasts of Africa and of North and South America.

In warm seas there exist a great number of species of different colored groupers. They are all members of the family Serranidae, which includes many fish: *Epinephelus, Cephalopholis,* and so forth, and especially those that belong to the genus *Stereolepis* or *Promicrops,* the latter reaching up to 10 feet and living off the west coast of Africa. *Grouper* is the general term for all fish of the Serranidae family.

HAWSER

A thick line used for mooring a ship.

HYDRACTINIA

A genus of the order Hydroida, class Hydrozoa, phylum Cnidaria. The Cnidaria include fixed species forming colonies and also these free-floating pelagic species, solitary and nonsolitary. Hydractinia commonly live on snail shells inhabited by hermit crabs.

KRILL

Euphausia superba, a small crustacean measuring two and one-third inches long. The green color that can be seen through the thin stomach wall is due to the algae it feeds on.

One of the constituent elements of plankton, this crustacean lives in cold waters. It is much more abundant in the Antarctic than in the Arctic. It multiplies with extraordinary speed during the Antarctic summer and covers the sea with a reddish-brown film. It is abundant to a depth of 30 feet, but can be found down to 3,000 feet. It reaches maturity two years after hatching. Krill is a major food for blue whales and finback whales as well as for various fish and birds.

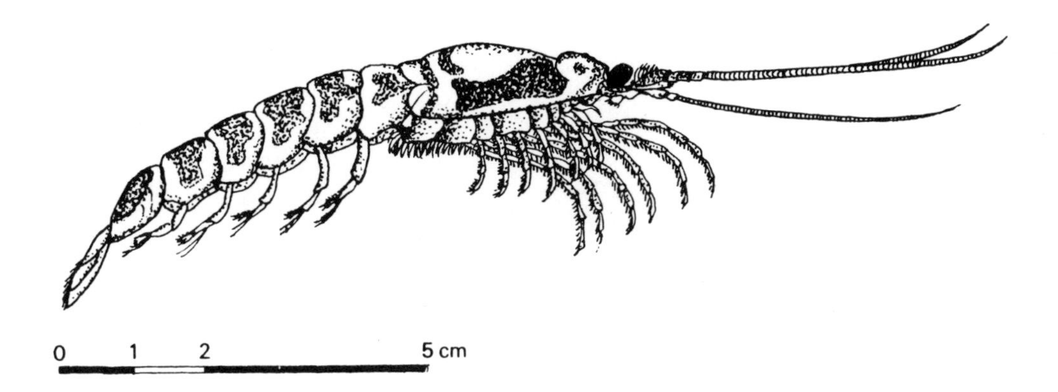

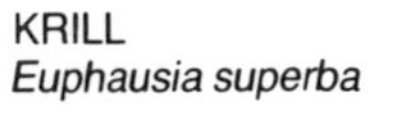

KRILL
Euphausia superba

A small crustacean that is the major food for blue whales.

LABORATORY BUOY

A floating laboratory conceived by J.Y. Cousteau, accommodating six persons: two crew members and four scientists. It was launched in the Mediterranean in January 1964 above a 7,500-foot sea bottom.

MERIDIAN

To take the meridian is to determine the latitude of a given spot by observing the meridian height of a star with a sextant.

NARCOSIS

A pathological stupor that can lead to loss of consciousness.

NARGHILE

Light diving gear manufactured by Spirotechnique. The diver is equipped with a regulator that is fed through a flexible tube from a source of compressed air on the surface.

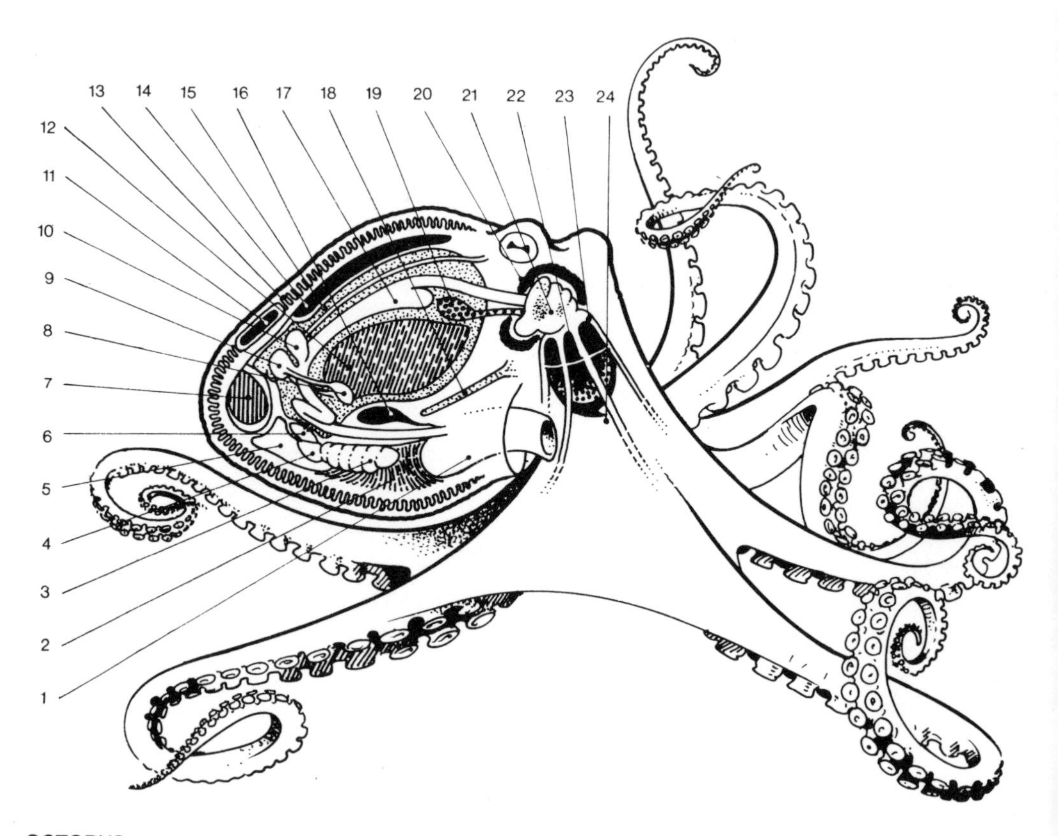

OCTOPUS
Octopus vulgaris

1. Funnel *2.* Muscles of the ventral cape *3.* Gills *4.* Gill heart *5.* Kidney *6.* "Systemic" heart *7.* Gonad *8.* Pancreas *9.* Cecum *10.* Stomach *11.* Rudimentary shell *12.* Liver *13.* Cape muscles *14.* Dorsal cavity of cape *15.* Intestinal space filled with blood *16.* Ink pouch *17.* Craw *18.* Cephalic vein *19.* Poison gland *20.* Cranian capsule *21.* Brain *22.* Mouth *23.* Arm nerve *24.* Beak

OCTOPUS

A mollusk of the class Cephalopoda, order Octopoda, genus *Octopus*. Octopi have twin gills and eight arms with suction cups. The third arm on the right, which is used in fertilizing, includes a channel to conduct the male reproductive elements. The best known of the octopi is *Octopus vulgaris*.

PENGUIN

A bird of the family Spheniscidae, order Sphenisciformes. There are 18 species, notably the Adélie penguin *(Pygoscelis Adeliae)*, the Gentoo penguin *(Pygoscelis papua)*, and the largest of all, the Emperor penguin *(Aptenodytes forsteri)*, which can reach three feet.

All these penguins are incapable of flight. Their anterior members are ailerons adapted to propulsion in water. They live in colonies, which can reach hundred of thousands of birds, in the Antarctic and on the southern coasts of South America, Australia, and New Zealand.

POMFRET

Chromis chromis of the family Pomacentridae is a small and very common fish along the Mediterranean coast. They are not frightened by divers, who often admire the beautiful light blue color of the young and the dark blue or brown silhouette of the adults. They do not grow over five inches.

POMPANO

A member of the Carangidae family, order Perciformes, close to the suborder Scombridea (which includes tuna and mackerel), pompanos are pelagic fish that live near coasts but that often are found in the open sea. There are numerous species, mainly in tropical seas.

They are very beautiful fish with a blue back with golden or silvery sides that resemble polished metal. Their lateral line is quite visible and their forked tail is carried on a very narrow stem. This makes it easy to recognize these fish even when swimming in tight schools in the transparent water of tropical lagoons.

RADIO FISH (MOORISH IDOL)

The divers of the Calypso call the *Zanclus canescens* by this name, although it is also referred to commonly as the Moorish Idol. It is a fish of coral seas with a thin body and a long dorsal fin that floats behind it, resembling a radio antenna. The Radio Fish is a member of the Acanthuridae family, order Perciformes.

ROCKFISH or SCORPION FISH

A variety of the family Scorpaenidae that measures up to 20 inches, the Rockfish is bright red or pink, with a large head covered with spines.

SALPA

A member of the order Salpidea, class Thaliacea, subphylum Tunicata, phylum Chordata. Found in all seas, the Salpa is especially found in warm or hot waters. The animal moves by contracting its muscular bands rhythmically and advances in a series of little hops. Its tunic is transparent but its colored viscera make a spot on the lower part of its body, the nucleus. That is where the heart and stomach are located. Their nervous system includes a large cerebroid ganglion and the sensory organs are composed of tactile lips and horseshoe-shaped eye.

SEA CUCUMBER

Phylum Echinodermata, class Holothuroidea. The Sea Cucumber is adapted to creeping. It has ventral and dorsal faces. The mouth is surrounded by ten branched tentacles. At the anus there are two large arborescent diverticles that serve as gills. There is an aquiferous apparatus for locomotion and two nervous chains. The Sea

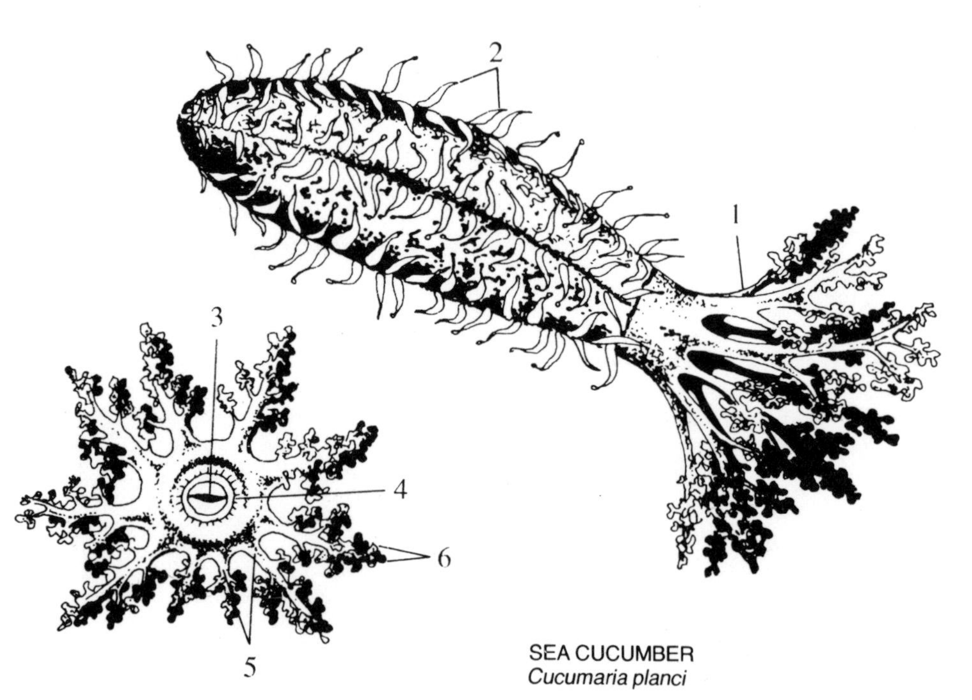

SEA CUCUMBER
Cucumaria planci

1. Tentacles
2. Ventral ambulacral feet
3. Mouth cavity
4. Circular lip
5. Large tentacles
6. Small tentacles

Cucumber feeds by swallowing sand in which it finds minuscule animal life. The sexes are separate, with males more numerous than females.

Sea Cucumber is a translation of its Latin name, *Cucumaria.* The Chinese attribute aphrodisiacal qualities to it.

SEAL

A marine mammal belonging to the order Pinnipedia and the family Phocidae. The Sea Leopard *(Hydruga leptonyx)* is one of the four species living in the Antarctic.

SPATANGINA or HEART URCHIN

A member of the order Spatangoidea, phylum Echinodermata with an oval or heart-shaped body covered with short spines. It lives in muddy sand and attains the size of a fist.

SPIDER CRAB

Common name of certain Decapodous crustaceans of the type Brachyura, the *Maja (Majidae)* characterized by a spiny body and long, thin appendages. These crabs carry algae in the rigid hairs on their shell.

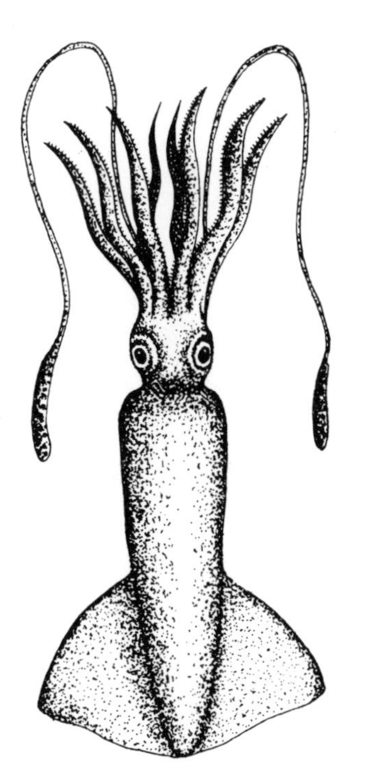

SQUID

SQUID

A member of the order Teuthoidea, class Cephalopoda, characterized by having ten tentacles and suction cups. The genus *Loligo* that is often eaten by man measures only about ten inches long. At certain seasons males and females crowd together in large groups for mating. This was the basis of one of the films of *The Undersea World of Jacques Cousteau : The Night of the Squid.*

The *Chiroteuthis,* with its slim body and very long tentacles, is a remarkable swimmer. The *Architeuthis,* the prey of sperm whales, is the giant of the group. This is the one that has inspired the stories about sea monsters such as sea serpents and the Norwegian sea monster Kraken.

Architeuthis' body may measure 20 feet long and the tentacles 41 feet. Little is known about this giant squid. It lives at depths of 10,000 to 13,000 feet and surfaces only at night. Its capture by man is rare and difficult.

STOCK

The transversal bar of an anchor. It can be fixed or moveable. Ancient anchors had a fixed lead stock, often very heavy and of large dimensions.

SURGEONFISH or TANG

Fish of the Acanthuridae family (order Perciformes) having an erectile spine on both sides of the base of the tail with which it can strike adversaries. It can inflict deep wounds, but they are not venomous. The spine can cut a diver's hand badly.

TRIGGERFISH

The Triggerfish (family Balistidae) owes its name to the first spine on its dorsal fin, which is blocked by a sort of safety catch. This enables the fish to lock itself in a coral hole where it is impossible to dislodge it.

TRUNK BUOY

A steel buoy supporting a chain attached to a weight on the bottom to which a ship can moor.

VIRGULARIA

A genus of Octocoralliaries, order Pennatulacea, class Anthozoa, phylum Cnidaria. Commonly known as Sea Pens, these animals can be found in certain depths of the Red Sea with white canes standing over three feet tall, forming virtual fields of growth.

photo credits

Les Requins Associés: pages 8, 12, 13, 34, 36, 37, 40, 41, 44, 48, 50, 51, 58, 62, 63, 64, 65, 69, 70 bottom, 73 bottom right, 76, 77, 80, 83, 85 bottom, 92, 93, 96 bottom two, 97, 100, 105, 107, 109 top, 113, 116, 117, 121, 124, 128, 133, 137, 140, 143, 144, 146, 147, 149, 150, 155, 158, 159, 161, 164, 169, 171, 173, 176, 179, 181, 182, 186, 187, 190, 193, 197, 200, 201, 204, 208, 212, 213, 217, 221, 229, 240, 243, 245 bottom left, 249.

Raymond Amaddio: pages 132, 232.

Albert Falco: pages 13, 16, 17, 20, 21, 28, 29, 32, 54, 55, 70 top, 73 bottom right, 85 top, 88, 90, 96 top, 109 bottom, 110, 136, 236, 237, 244, 245 top left and right and bottom right, 250, 251, 252, 254, 256.

Paul Zuena: pages 224, 228.

Jacques Roux: page 225.

Drawings in the appendix and glossary are by Jean-Charles Roux.

index

(page numbers in italic indicate photos)

Dépôt légal 1er trimestre 1976 – Flammarion, éditeur, N° 9494 – N° d'imp. : 5930
Imprimerie Déchaux, Aulnay-sous-Bois